AF552751

About the Authors

Manju Prem S. is a Ph.D. candidate in Agriculture Extension at Kerala Agricultural University and a UGC NET-Junior Research Fellow in two subjects namely 'Adult and Continuing Education and Extension' and 'Education'. With an MSc from Bidhan Chandra Krishi Viswavidyalaya and a BSc from the University of Agricultural Sciences, Bangalore, he has a strong foundation in agricultural sciences. His research and publications focus on climate resilience in agriculture, and he has participated in numerous international conferences and seminars. He has also published a book on adult and continuing education and extension. His expertise and contributions make him a valuable resource in the field of agricultural extension.

Arindam Deb is a dual Ph.D. candidate at Kerala Agricultural University and Western Sydney University. He has qualified for the National Eligibility Test in two subjects: Environmental Sciences (UGC) and Agronomy (ICAR). He holds a B.Sc. in Agriculture from Assam Agricultural University and a M.Sc. in Agronomy from Central Agricultural University. His doctoral research focuses on the effects of biotic and abiotic stresses on rice under the scenario of climate change.

Praveen Gidagiri is a Ph.D. Candidate in the Department of Postharvest Management at Kerala Agricultural University. He is a graduate and postgraduate from the University of Horticultural Sciences, Bagalkot, Karnataka. He qualified for ICAR NET in 2019. He has published 4 research papers, 12 popular articles, 1 book chapter, and presented papers in national and international conferences. He was also awarded the Best Oral Presentation award and the Best M.Sc. Thesis award. He has experience of more than 2 years in teaching. His area of specialization is postharvest management of horticultural crops.

Mitigation and Adaptation Strategies for Climate Change

Manju Prem S.
Arindam Deb
Praveen Gidagiri

1168, Sector 13, Urban Estate
Karnal-132 001, Haryana
Tel: 91-84470 75807, 18440 41168
Email: contentvibesppa@gmail.com
www.contentvibes.in

Print ISBN: 978-81-977815-5-1

ebook ISBN: 978-81-977815-4-4

Preface

This book is the result of a conversation among peers about climate change mitigation strategies in various disciplines within agriculture. We found the discussion intriguing and realized that such comprehensive literature was lacking. Thus, we conceptualized this book to bridge that gap and serve as a souvenir for our Ph.D. batch.

Our primary goal was to gather information on the effects of climate change on agriculture and its mitigation strategies. The book provides detailed insights into how different disciplines tackle the threat of climate change. While this is an earnest attempt, we acknowledge that it is impossible to cover the entirety of such a vast subject in a single book. We welcome constructive criticism and suggestions from respected teachers, students, and all concerned individuals, as this will help us improve and refine the book in the future.

We extend our deepest gratitude to the researchers who have dedicated their lives to understanding the effects of climate change. Their invaluable contributions form the foundation of our work. Additionally, we thank our peers for their collaboration, support, and insightful discussions that have greatly enriched this book.

Editors

Contents

1

Mitigation and Adaptation Strategies for Climate Change and Agronomic Properties

Merin Jose, Navya M V, Greeshma U and Arindam Deb

College of Agriculture, Kerala Agricultural University. Vellayani-695522

Abstract

Climate change denotes long term variations in weather parameters and has direct influence on crop growth and productivity. Raise in temperature, irregular rainfall and extreme climate are associated with climate change and has several effects on crops that have been discussed in the chapter. Judicious agronomic crop management practices has the ability to alleviate the effect of climate change and make the farming system climate resilient. Concepts such as conservation agriculture, maintaining agroecology, agroforestry, climate resilient agriculture have been elaborated in this chapter.

Keywords: *Climate change, conservation agriculture, agroecology, agroforestry, climate resilient agriculture.*

Introduction

The foundation of agricultural productivity lies in the realm of weather and climate, a reality unaltered by technological progress. The consequences of climate change on agriculture represent a worldwide apprehension, particularly significant for India. Around 54.6% of the Indian population is engaged in agriculture and allied activities, contributing to 17.4% of the country's total gross value added. The agriculture sector is battling a multitude of challenges, such as increasing food demand, depletion of agricultural lands and water resources, and the devastating effects of climate change. It is high time that we acknowledge and address these issues to safeguard the future of our food systems and ensure food security for all.

Climate change refers to the long-term variations in weather patterns and temperature over extended periods. Climate change occurs when earth's climate system experiences variations in weather patterns that can last for decades or millions of years. These changes can arise naturally due to phenomena like volcanic eruptions or changes in solar activity. However, since the 1800s, human activities accelerate the natural pace of climate change, primarily due to the combustion of fossil fuels like coal, oil, and gas. These activities release greenhouse gases like carbon dioxide (CO_2), methane (CH_4), and nitrous oxide (N_2O) into the atmosphere, trapping heat and leading to a gradual increase in global temperatures. One of the most remarkable characteristics of climate change is the increase in temperature, which is widely referred to as global warming (Houghton, 2009). According to the report of Intergovernmental Panel for Climate Change, the release of greenhouse gases caused by human actions has contributed to around 1.1°C of warming since the period between1850-1900. The report also predicts that the global temperature will likely increase to 1.5°C or more in the next 20 years (IPCC, 2021).

Climate change is a globally recognized environmental problem and, global assessments have classified South Asia as the region that is particularly vulnerable to impacts of climate variability and change (Nagabhatla *et al.*, 2015). Climate change has far-reaching consequences that affect multiple facets of the environment and society and include intense droughts, water scarcity, severe fires, rising sea levels, flooding, melting polar ice and declining biodiversity.

Impacts of climate change on agriculture

Climate change is having a significant impact on global food production, affecting both the supply and demand of food grains around the world (Shanker *et al.*, 2018).India has 17 per cent of the world's population, with only 4 per cent of usable water resources (Jain *et al.*, 2019). The agricultural sector alone utilizes 80 per cent of groundwater resources. (Harsha, 2017). Meeting the increasing and diverse demands of a growing population while also managing dwindling and valuable natural resources, global agriculture faces significant challenges. Studying the potential impacts of climate change on this sector is crucial for ensuring its sustainability. The consequences of climate change are far-reaching and include:

1. Increasing temperature

The Earth absorbs approximately 75% of the solar energy reaching the surface, consequently raising its temperature. A portion of this absorbed energy is radiated back into the atmosphere. Certain gases present in the atmosphere, such as ozone, methane, carbon dioxide, water vapor, and chlorofluorocarbons, are referred to as greenhouse gases. They trap some of this heat, preventing it from escaping into space. These gases contribute to the warming of the atmosphere, leading to global warming.

The 2023 Global Climate Report by NOAA National Centers for Environmental Information revealed that each month of 2023 ranked within the top seven warmest for their respective months. Furthermore, the months from June to December were the hottest on record. In July, August, and September, global temperatures exceeded 1.0°C (1.8°F) above the long-term average, marking the first time in NOAA's records that any month surpassed this threshold.

2. Irregular rainfall

Water plays a crucial role in plant growth. Fluctuations in precipitation patterns have a profound effect on agriculture. Solar radiation serves as the primary energy source driving the water cycle. The increase in greenhouse gas concentrations caused by human activities has contributed to a net rise in radiation input. Consequently, elevated global temperatures result in increased evaporation, thereby accelerating the hydrological cycle. This increased evaporation gives rise to more cloud formation and precipitation in the form of rain. As a result, rainfall patterns and intensity undergo alteration. Indian Institute of Tropical Meteorology (IITM) Pune highlighted that for every 1°C increase in temperature, the atmosphere can retain 7% more moisture. Additionally, excessive rainfall is due to the rapid warming of the world's oceans, which have absorbed 90% of the additional heat generated by human-induced climate change over the past five decades. This phenomenon not only contributes to increased instances of flooding due to excess rainfall but also amplifies the severity and recurrence of droughts in regions already susceptible to arid conditions.

Climate change is significantly altering the monsoon patterns. Rainfall is becoming increasingly unpredictable, characterized by more frequent heavy rainfall events with long dry spells in between, along with notable regional disparities in rainfall distribution. However, water availability is not solely determined by precipitation. The escalating evaporative demand due to higher temperatures and extended growing seasons could potentially elevate global crop irrigation needs by 5 to 20 percent, or even more, by the 2070s or 2080s.

3. Effect on crops

The variation in monsoon patterns has also affected cropping systems in the Indian subcontinent, where the primary crop planting seasons are kharif (beginning with the onset of the monsoon) and rabi (planted after the monsoon). The Kharif season heavily depends on the monsoon, and changes in the timing of monsoon onset and withdrawal impact both cropping seasons. Heavy rainfall in September reduces yields of short-duration kharif crops such as groundnut, black gram, soybean, and maize and also affects storage and transportation, leading to food inflation and a food crisis. Irregular rainfall coupled with rising temperatures and humidity result in increased pest infestations and disease outbreaks (Gupta *et al.*, 2023). This, in turn, can adversely affect the quality and nutritional value of grains. Ye *et al.* (2013) observed a significant reduction in paddy yields under alternate wetting and drying irrigation system when compared to continuous flooding irrigation.

Extreme heat stress, typically exceeding 40°C, can inflict damage upon plants. The severity of the impact intensifies with the degree and duration of exposure. Heat stress can lead to various detrimental effects on plants, including hindered seed germination, stunted growth, improper development, changes in photosynthesis, phenology, and dry matter allocation, increased water loss, reduced yield, and diminished crop quality. Heat stress, indicated by temperatures exceeding 30°C, can halt pollen development during anthesis, hindering the transfer of resources to developing grains, as observed in wheat (Zahra *et al.*, 2023). Drought conditions diminish plant transpiration rates, potentially leading to leaf rolling and desiccation, reduced rates of leaf expansion and overall plant biomass, and immobilization of solutes, exacerbating heat stress on leaves. Furthermore, disrupted leaf water relations and root hydraulic conductivity have been documented in tomato (Morales *et al.*, 2003). High temperatures cause plants to lose water more rapidly during the day time than at night. Zhao and Fitzgerald (2013) found that rising temperatures increased respiration, triggering an uptick in carbon metabolism but causing a decline in paddy yield. Similarly, Asseng *et al.* (2011) demonstrated that global warming may diminish net carbon gain by elevating plant respiration rates, consequently reducing crop yields and potentially fostering the proliferation of weeds, pathogens, and pests. Gammans *et al.* (2017) conducted an analysis of winter wheat yields in France, employing a modeling technique that utilized historical yield data and gridded weather information spanning from 1950 to 2015. Their model projected that as a consequence of climate change, there would be a decline in wheat yield ranging from 3.5% to 12.9% in the medium term, spanning from 2037 to 2065. Additionally, it forecasted a more substantial reduction of 14.6% to 17.2% in winter wheat production by the end of the century.

Agriculture relies heavily on climate conditions, making it vulnerable to the effects of climate change. Therefore, it is imperative to implement adaptation strategies to mitigate its impact on agriculture.

Crop management practices for climate resilience

The adverse effects of climate change in Agriculture are of serious concern as it can jeopardize global food security. Therefore, adopting efficient crop management practices capable of enhancing productivity whilst being resilient to changing climate is the need of the hour. The crop management practices must be sustainable and capable of mitigating climate change. Considering the above points lets discuss the sustainable crop management practices for crop resilience:

1. **Selection of crops:** The choice of crops is the first and foremost decision taken by the farmer. This decision is always made considering climate, soil, market and government policies. Another aspect that must be taken into consideration for the same is climate change. Again, the crops that are grown in several parts of India have cultural ties and are integral part of diet, for example, rice is a predominant crop in the eastern part of the country and wheat in the Gangetic plains. Hence, the selection of crops must be done keeping in mind the predominant crop in the region, and the selected crops should complement the major crop in the area. There comes the importance of diversification of crops. Diversification can be achieved in two ways: spatial and temporal. All kinds of crop rotation come under temporal diversification and intercropping comes under spatial diversification. Increasing the number of crops both temporally and spatially leads to more biomass production and thus enhances the carbon sequestration in agricultural systems. However, such increase comes with increase in the cost of inputs like nutrients and water. The most basic principle of diversification is the inclusion of legumes. Addition of dry peas in crop rotations enhances the yield of cereals and oilseeds while reducing N-fertilizer requirement. Pulses being deep rooted crops can utilize unused water from deeper layers thus enhancing overall water use efficiency (WUE) of the system (Liu *et al.*, 2022). Another important component that needs serious considerations while selecting a cropping system for climate resilience is millets. Millets being domesticated in semi-arid regions of Sub-Saharan Africa and Asia possesses inherent tolerance to sub optimal climatic conditions (Wilson and VanBuren, 2022).

2. **Selection of varieties:** Variety choice is an important step towards climate resilience as its potential of rapid implementation. The results of proper variety selection can be seen within one cropping season. Heat and drought stress is associated with climate change and several tolerant varieties of cereals, pulses and vegetables are available. Sahbhagidhan is a drought tolerance rice variety and IRRI claims drought tolerant rice varieties have yield advantage of 0.8 to 1.2 tonnes per hectare over susceptible ones under drought conditions (IRRI, 2024). Though stress tolerant varieties are available for almost all cultivated crops, there is still huge scope for improvement. Around 7.4 million of landraces, cultivars and wild relatives of major crops are conserved in more than 1750 gene banks which can be used to churn out traits to improve resilience of crops (Benitez-Alfonso *et al.,* 2023).

3. **Time of sowing:** Time of sowing is one of the most important non-monetary inputs in agriculture. Climate change brings with it aberrant rainfall and unusually high temperatures. Higher temperature during vegetative stage can adversely impact the harvest index of wheat (Hussain *et al.,* 2018). However, these adverse effects can be mitigated using agronomic adaptations such as temporal variation of sowing dates. Based on the meteorological forecast, sowing dates of crops can be modified so as to avoid harsh temperature and drought during critical stages of crop growth.

4. **Optimum plant population:** Plant population or density influences all the physiological and yield parameters of a crop. Excessive plant population leads to competition with the crop and thus leads to wastage of resources. While lower plant population leaves excessive space in the field leading to weed growth and high evaporative losses of water. There for maximization of yield optimization of plant population is important.

5. **Nutrient management:** Nutrient management is a crucial aspect of crop production as it helps to maintain soil health that in turn enhances sustainable productivity. A sound nutrient management strategy is a must for a climate resilient crop production system. Improper nutrient management leads to wastage of nutrients, pollution of groundwater, river bodies and greenhouse gas emissions. Therefore, for judicious use of resources the concept of integrated nutrient management (INM) is important. INM refers to combining all the available sources nutrients (organic and inorganic) in such a way that they complement each other, reduce the external input requirement, make the system ecologically sound and economically viable. It emphasizes the importance of soil

health, thus addition of organic matter is an important aspect of INM. This in turn helps in better recycling of nutrients by synchronizing nutrient demand and release (Wu and Ma, 2015).

6. **Water management:** Agriculture is the principal consumer of global ground water and with variability in climatic conditions it is estimated to increase by 14.7 % by 2095. Thus, there is a need to efficiently manage water and reduce wastage. It is always preferred to use efficient irrigation methods like drip and sprinkler wherever possible. In crops such as rice, where water consumption is huge and micro irrigation is not feasible, several tried and tested techniques to save water exists, for eg, alternative wetting and drying, SRI method, upland and aerobic rice etc. Other options to save water include mulching, growing of cover crops, crop diversification, choice of drought tolerant crops, etc.

7. **Pest management:** Climate change in last three decades has influenced interactions between plants and insect pests (Castex *et al.,* 2018). Such changes can lead to emergence of new pests or make current pest problems more severe. Moreover, changing climate can led to failure of various biological control of insects and weeds. Therefore, there is a need to understand the dynamics interactions between plants and insects, so as to come up with decision support systems to curtail the damage to crops due to biotic stress.

Conservation Agriculture (CA): It is a farming practice that ensures minimum soil disturbance, permanent soil cover and diversification of plant species. Tillage operations are mostly avoided which might lead to increase in weed population for the initial years, however as the seasons will pass the seedbank of weeds in the top layer of soil will reduce with proper weed management leading to lesser loss of nutrients and water by weeds. Permanent soil cover in CA is done by retaining 30% crop residues, which slows the decomposition compared to incorporation of residues, thereby increasing carbon sequestration in soil. Conservation Agriculture requires 20 to 50% less labor, leading to a decrease in greenhouse gas emissions due to reduced energy inputs and enhanced nutrient use efficiency. Simultaneously, it safeguards and preserves the soil, preventing its degradation and the release of carbon into the atmosphere (FAO, 2024).

Agroecology and its importance in climate change

Agroecology is an integrated approach that takes into account ecological and social concepts and principles to design and manage sustainable farming system. It aims to optimize the relationships that exist between humans, animals, plants, and the environment while also considering the social issues that must be addressed for a productive and sustainable food system (FAO, 2018).

Agroecological practices like farm diversification, organic agriculture and agroforestry can make a significant contribution in helping low and middle-income countries in achieving climate adaptation and mitigation targets through their food systems. Different principles of agroecology in compacting climate change (HLPE, 2019) can be discussed as follows:

- **Diversity**: Diversity in agroecological systems includes crop diversification, preservation of local genetic diversity, animal integration, management of soil organic matter, water conservation and harvesting, and livelihood diversification that reduces vulnerabilities to climate variability is fundamental in addressing climate change.
- **Synergies**: Agroecological practices strive to enhance favorable ecological interaction, synergy, integration, and complementarity among the elements of agroecosystems including plants, animals, trees, soil, water providing opportunities to create synergies and balance trade-offs across the multiple goals of food security and climate change adaptation and mitigation.
- **Efficiency**: This principle focuses on gradual reduction in the use of pesticides and synthetic fertilizers, which are replaced by biological methods. Such a substitution avoids the climate damaging emissions that arise when these substances are produced and used.
- **Resilience**: It is a fundamental outcome of agroecological systems based on the implementation of a various other principles, notably the spatial and temporal diversity, alongside all the traditional knowledge of smallholders, family farmers, or indigenous people and their associated social networks that enabled them to effectively manage climate risks, both past and present.
- **Recycling**: The recycling of biomass, with an objective of optimizing organic matter decomposition and nutrient cycling over time plays a crucial role in improving the efficiency of resource utilization. This practice not only reduces waste and production costs but also contributes to limiting greenhouse gas emissions. Recycling offers multiple benefits by closing nutrient cycles and reducing waste. Recycling also permits producers to save costs on inputs, reducing their vulnerability to price volatility and climate related shocks.

- **Co-creation and sharing of knowledge**: Traditional knowledge is seen as the foundation for current and forthcoming agricultural innovations and technologies to deal with climate change. Despite the need for new knowledge, particular emphasis is placed on addressing power imbalances between scientists and farmers, recognizing the historical effectiveness of traditional farming methods in overcoming climate challenges.
- **Human and social values**: Agroecology focuses on addressing social constraints and power dynamics. Through fostering autonomy and enhancing adaptive capabilities in managing agro-ecosystems, agro ecological practices empower individuals and communities to take control of their circumstances and tackle challenges such as poverty, hunger, malnutrition, and climate change.
- **Culture and food traditions**: Agroecological systems are rooted in the culture, identity, and traditions of local communities, which promote culturally appropriate, healthy, diverse, and seasonally suitable diets. These foundations serve as the basis for developing innovative systems capable of adapting to climate change.
- **Responsible governance**: This principle aims at establishing supportive national and local frameworks that reduce the hindrances at local level, recognize and support the needs and interests of family farmers, smallholders, in order to avoid maladaptation to climate change.
- **Circular and solidarity economy**: This principle seeks to re-establish links between producers and consumers, offering creative solutions for living within the limits of our planet while also fostering a socially inclusive foundation for sustainable development. Specifically, it involves advocating for fair solutions tailored to local needs, resources, and capabilities, thus fostering more equitable and sustainable markets.

Potential of agroforestry as a climate change mitigation strategy

Agroforestry is a method of land usage that is environmentally friendly and increases overall productivity by growing food crops alongside trees, livestock, and/or other crops on the same plot of land. It can be characterized as an effective and integrated land use management system that involves growing specific agricultural crops, forest tree species, and/or animals on the same plot of land in a sequential or simultaneous manner while using suitable management techniques. This leads to an overall increase in production under specific climatic and edaphic conditions and takes into account the socioeconomic status of the local population (King, 1969).

As emphasized in the reports by the Intergovernmental Panel on Climate Change (IPCC), agroforestry is a promising agroecological approach to climate change adaptation due to its multitude of co-benefits that many agroforestry systems provide in addition to climate change adaptation, including synergies with climate change mitigation through carbon sequestration, enhanced food security and income opportunities, the provision of ecosystem services, and biodiversity conservation (IPCC, 2007).

The tree component of agroforestry is principally responsible for its significant contribution to mitigating climate change. The trees over course of time sequestrate huge quantity of carbon in their biomass. It facilitates both adaptation to and mitigation of climate change. It can also help to boost food production and provide alternative sources of nutrition or income when crop yields are low. With climate change expected to lead to unpredictable seasons in the future, placing even greater pressure on agricultural systems, food production and food prices, agroforestry is a viable option to help buffer farmers against the impacts.

The general biophysical benefits of agroforestry systems are well understood at both the farm and landscape level (Lasco *et al.*, 2014). Agroforestry can create microclimates with lower mean air temperatures, reduce crop transpiration rates by shading crops, draw water from deeper soil layers and support root water uptake by crops, minimize soil loss during heavy rainfall or downstream flood events and create windbreaks and buffer crops from storms. Agroforestry can increase the resilience of agricultural soils by increasing soil nitrogen and carbon, as well as increasing spatial heterogeneity of the soil microbial community (Guillot *et al.*, 2019).

The socioeconomic benefits of agroforestry systems are well known and encompass the provision of food and income derived from trees, crops, and livestock. Recent studies indicate that these socioeconomic benefits can help farmers adapt to climate-induced extreme events by increasing their adaptive capacity and reducing their vulnerabilities. In Kenya, farmers have demonstrated how they are able to increase resilience to drought through the planting of drought-tolerant tree species (Quandt, 2020). Additionally, agroforestry systems create diversified sources of income for farmers, adding to their ability to cope with extreme weather events. In addition, reforestation, agroforestry, mixed farming, and the adoption of resistant plant varieties are important corrective measures to lessen climate change and improve people's socioeconomic standing.

Hence, agroforestry emerges as a promising adaptation strategy, aligning with the principles of Climate Smart Agriculture, particularly beneficial for smallholder farmers across the developing world.

Progressive transformation: shifting towards climate smart agriculture and climate resilient agriculture

Climate change affects agriculture differently across regions and time periods, with outcomes being diverse and uncertain. The evolution of agriculture in the future will depend on how it responds to these changes. Under this scenario farmers need to adjust and adapt their practices to suit changing climatic conditions, and agricultural activities will need to be modified to reduce emission of greenhouse gas. Agricultural sector is at the forefront of the climate crisis in terms of being a significant source for greenhouse gas emissions while also being negatively impacted by rising temperatures, droughts and floods. Much agricultural research aims to mitigate the impacts of climate-related risk and to improve resilience in the face of climate variability and extreme weather conditions. Climate-smart agriculture (CSA) and climate-resilient agriculture (CRA) are two strategies aimed at addressing the challenges posed by climate change to food security and the sustainability of agriculture (Hellin *et al.*, 2023).

CSA involves a wide range of agricultural technologies and practices, including, conservation agriculture, stress-adapted crop germplasm e.g., drought-tolerant maize varieties, agroforestry, and soil and water conservation (Lipper *et al.*, 2017). It is defined as a strategy to tackle the challenges posed by climate change and food security by sustainably improving productivity, building resilience, reducing GHG emissions, and enhancing achievement of national food security and development goals (FAO, 2010). CSA aims to foster sustainable agricultural development that adapts to climate change while contributing to mitigation efforts.

CSA aims to simultaneously achieve three goals: (Lipper *et al.,* 2017)

- **Enhanced Productivity**: CSA practices are designed to enhance agricultural productivity and production systems, ensuring food security for a growing global population. This involves adopting innovative techniques and technologies that optimize resource use and maximize yields.
- **Climate Change Adaptation**: CSA prioritizes resilience-building strategies to help farmers and ecosystems adapt to the impacts of climate change. This includes implementing practices such as crop

diversification, water management, and soil conservation to mitigate risks associated with extreme weather events, changing precipitation patterns, and temperature fluctuations.

- **Mitigation of Greenhouse Gas Emissions**: Recognizing the role of agriculture in contributing to greenhouse gas emissions, CSA emphasizes mitigation strategies to reduce the carbon footprint of agricultural activities. This may involve practices such as agroforestry, conservation tillage, and improved livestock management to sequester carbon and minimize emissions.

Climate-resilient agriculture (CRA) is a subset of climate smart agriculture interventions that specifically targets, enhancing the resilience of agricultural systems and the associated social systems. It encompasses adaptation and mitigation strategies, along with the effective utilization of biodiversity across various levels, including genes, species, and ecosystems. Thus, CRA is considered a crucial prerequisite for sustainable development amidst changing climatic conditions. Recognizing the current impacts of climate change on agricultural systems and livelihoods, climate-resilient agriculture aims to proactively strengthen resilience, reduce vulnerabilities, and enhance adaptive capacity. By utilizing a mix of location-specific strategies and capitalizing on local knowledge and resources, climate-resilient agriculture endeavors to develop more robust and sustainable food systems capable of enduring future climate challenges (Rao *et al.*, 2019).

Key components of CRA include (Tripathi and Bisen, 2019):

- **Adaptive Crop Management**: Implementing crop varieties and management practices that are resilient to climate variability, such as drought-tolerant crops, flood-resistant varieties, and improved water management techniques.
- **Soil Health Improvement**: Enhancing soil fertility, structure, and moisture retention capacity to mitigate the impacts of climate change on crop production. This may involve practices like conservation tillage, cover cropping, and organic matter incorporation.
- **Water Resource Management**: Promoting efficient water use through techniques like rainwater harvesting, drip irrigation, and integrated water management systems to cope with changing precipitation patterns and water scarcity.
- **Livelihood Diversification**: Encouraging farmers to diversify their income sources and adopt alternative livelihood strategies to reduce

dependence on climate-sensitive activities and enhance resilience to climate risks.

While there is overlap between the concepts of CSA and climate-resilient agriculture, CSA encompasses a wider range of objectives, including mitigation of greenhouse gas emissions, whereas climate-resilient agriculture primarily concentrates on building resilience to climate change impacts. Both approaches are vital for ensuring the sustainability and stability of global food systems in the face of climate change.

India has good reason to be concerned about the negative effects of climate change on its economy. As a developing nation, its population is heavily dependent on sectors that are sensitive to climate change, making it extremely vulnerable to climate change. Climate change can have serious impact on its crops, forests, coastal regions, etc. which may have an impact on the accomplishment of crucial developmental goals.

Various Steps and Policies Adopted by India to Mitigate Climate Change

India has a comprehensive system of institutional and legislative processes in the area to address the severe environmental problems it is facing due to urbanization, industrial expansion, and population growth. India is among the developing countries that have included specific environmental protection measures in its Constitution.

The Stockholm Conference and increasing awareness about the environmental crises prompted the Indian Government to enact the 42nd Amendment to the constitution in 1976. The amendment introduced direct provisions for the protection of the environment by adding Article 48-A to the Directive Principles of State Policy.

ARTICLE 48A

Article 48A of the Constitution of India provides that 'the State shall endeavor to protect and improve the environment and to safeguard the forests and wild life of the country'.

ARTICLE 51A (g)

It deals with the fundamental duties of the individuals and states that "It shall be the duty of every citizen of India to protect and improve the natural environment including forests, lakes, rivers and wildlife and to have compassion for living creatures."

ARTICLE 253

It states that "Parliament has the power to make any law for the whole or any part of the country for implementing any treaty, agreement or convention with any other country."

The regulatory and institutional decision-making framework for environment protection is embodied in nine major acts of the Indian Parliament. They are as follows:

- Water Act (Prevention and Control of Pollution), 1974; Water Cess Act, 1974
- Air (Prevention and Control of Pollution) Act, 1977
- Forest Conservation Act, 1980 (amended in 1988)
- Environment Act, 1986
- Motor Vehicle Act, 1988
- Public Liability Insurance Act, 1991
- Notification on the Coastal Regulation Zone, 1991
- National Environment Appellate Authority Act, 1997)
- The Plastic Waste Management Rules, 2016
- The Solid Waste Management Rules, 2016

Case Studies

1. Addressing Drought Vulnerability by Cultivation of Aerobic Paddy (MAS-26)

The demonstration of aerobic paddy MAS-26 was conducted at farmer field Sri. Mahesh. N. M, in 0.5 ha. This method of cultivation was adopted in Tumkur district in Karnataka because in traditional rice cultivation method large quantities of water is required and it is highly labor intensive. The main advantages of adopting this drought tolerant aerobic paddy MAS 26 are –

- Direct sowing
- No need for puddling
- Resistance to pests and diseases
- Reduction in pollution
- Medium duration

- Possibility of about 60 tillers on an average per seed
- 50% water saving along with 80% seed saving

The result showed that the performance of the MAS-26 (37.5 q/ha) was found to be superior to the local cultivated variety of paddy (29.8 q/ha). The yield of MAS-26 increased 12.4 % and the farmer got advantage with and extra income and yield of Rs. 3,600 and 2 q ha^{-1} respectively.

2. Innovative Vegetable Women Farmer – a Success Story through adoption of new technologies

Smt. Shashikala is an innovative vegetable farmer from Pemmanahalli village of Udigere Hobli, Tumakuru Taluk and Tumakuru District, Karnataka. She started growing vegetables like french beans, tomato, peas, brinjal, chilli and green leafy vegetables since 2010. Earlier they used to grow field crops like ragi, field bean, red gram and jowar. She used improved peas variety of IIHR (Arka Ajit) and Magadi local for cultivation to overcome the climate change effects like pest and diseases. By adopting various high yielding varieties, average net returns of Rs. 20,000/- per acre was obtained and also switching over to short duration vegetable crops, she is able to earn a net return of Rs. 60,000/- per acre.

3. In Jamaica, farmers struggle to contend with a changing climate

Farmers in Jamaica, an island nation of 3 million, are especially vulnerable to changing climate. In 2020, Jamaica became the first Caribbean country to submit a tougher climate action plan to the UN because the country was at risk from rising sea levels, drought and more intense hurricanes, its government said. The major risks in cultivation of crops were droughts, floods, and the spread of pests, which are the byproducts of climate change and also threatening agricultural production around the globe.

In 2018, the Mount Airy farmers enrolled in the United Nations-backed programme that helps build the resilience of communities to threats such as climate change, poverty and water insecurity. It is regarded as the first joint programme of the United Nations in Jamaica, combining the resources of six agencies, including UNEP. In Mount Airy, the UN programme has invested in 30 new water harvesting systems. The large, black tanks, which appear across the hilltops like turrets, catch and store rainfall, allowing the farmers to use it evenly through drip irrigation system. This reduces the emerging threat of longer and more intense dry spells.

The new irrigation system helps the farmers a lot by making farmers free from watering their crops by hand and using this system they can mix fertilizer with water and spread it evenly among the crops, saving the farmers valuable time. The dissolvable fertilizer is also cheaper than standard fertilizers.

Conclusion

Climate change has adverse effects on crop growth and may pose severe threat to global food security. However, a glimmer of hope is seen as various agronomic management practices can curtail the impact of climate change on agricultural system. Various governmental policies are being implemented to mitigate climate change. Several case studies provided shows that adoption of new technologies can help in sustaining the production. Therefore, though we are facing a critical problem, with an all-round effort from farmers, government, and researchers we can tackle climate change successfully.

References

Asseng, S., Foster, I. and Turner, N.C. 2011. The impact of temperature variability on wheat yields. Glob. Chang. Biol. 17(2): 997–1012.

Benitez-Alfonso, Y., Soanes, B.K., Zimba, S., Sinanaj, B., German, L., Sharma, V., Bohra, A., Kolesnikova, A., Dunn, J.A., Martin, A.C. and u Rahman, M.K., 2023. Enhancing climate change resilience in agricultural crops. Current biology, 33(23), pp.R1246-R1261.

Castex, V., Beniston, M., Calanca, P., Fleury, D. and Moreau, J., 2018. Pest management under climate change: The importance of understanding tritrophic relations. Science of the Total Environment, 616, pp.397-407.

FAO, 2024. Food and agriculture organisation. Available at: https://www.fao.org/conservation-agriculture/overview/why-we-do-it/en/#:~:text=Conservation%20Agriculture%20is%2020%20to,releasing%20carbon%20to%20the%20atmosphere. Accessed on 2/3/2024.

FAO [Food and Agriculture Organization]. 2010. Climate smart agriculture: policies, practices and financing for food security, adaptation and mitigation. Food and Agriculture Organization. Rome, Italy.

FAO [Food and Agriculture Organization]. 2018. The 10 elements of agroecology: Guiding the transition to sustainable food and agricultural systems. Food and Agriculture Organization. Rome, Italy.

Gammans, M., Merel, P. and Ortiz-Bobea, A. 2017. Negative impacts of climate change on cereal yields: Statistical evidence from France. Environ. Res. Lett. 12(5): 54007p.

Guillot, E., Hinsinger, P., Dufour, L., Roy, J. and Bertrand, I. 2019. With or without trees: resistance and resilience of soil microbial communities to drought and heat stress in a Mediterranean agroforestry system. Soil Biology and Biochemistry, 1291: 22-135.

Gupta, A.K., Agrawal, M., Yadav, A., Yadav, H., Mishra, G., Gupta, R., Singh, A., Singh, J. and Singh, P. 2023. The effect of climate change on wheat production: Present patterns and upcoming difficulties. Int. J. Environ. Clim. Chang. 13(11): 4408-4417.

Harsha, J. 2017. Micro-irrigation in India: an assessment of bottlenecks and realities. Available:http://www.globalwaterforum.org/2017/06/13/micro-irrigation-inindian-assessment-of bottlenecks-and-realities.

Hellin, J., Fisher, E., Taylor, M., Bhasme, S. and Loboguerrero, A.M. 2023. Transformative adaptation: from climate-smart to climate-resilient agriculture. CABI Agriculture and Bioscience, 4(1), 30p.

HLPE. 2019. Agroecological and other innovative approaches for sustainable agriculture and food systems that enhance food security and nutrition. A report by the High-Level Panel of Experts on Food Security and Nutrition of the Committee on World Food Security, Rome, 163p.

Houghton, J., 2009. Global warming: the complete briefing. Cambridge university press, 430pp.

Hussain, J., Khaliq, T., Ahmad, A., Akhter, J. and Asseng, S., 2018. Wheat responses to climate change and its adaptations: a focus on arid and semi-arid environment. International Journal of Environmental Research, 12, pp.117-126.

IPCC 2021. Summary for policymakers. In V. Masson-Delmotte, P. Zhai, A. Pirani, S. L. Connors, C. Péan, S. Berger, N. Caud, Y. Chen, L. Goldfarb, M. I. Gomis, M. Huang, K. Leitzell, E. Lonnoy, J. B. R. Matthews, T. K. Maycock, T. Waterfield, O. Yelekçi, R. Yu, & B. Zhou (Eds.), Climate change 2021: The physical science basis. Contribution of working group I to the sixth assessment report of the Intergovernmental Panel on Climate Change. Cambridge University Press.

IPCC [Intergovernmental Panel on Climate Change]. 2007. Synthesis report summary for policymakers. An Assessment of the Intergovernmental Panel on Climate Change, 7-22p.

IRRI, 2024. Available at https://www.irri.org/climate-change-ready-rice, accessed on 1/3/2024

Jain, R., Kishore, P. and Singh, D.K. 2019. Irrigation in India: status, challenges and options. J. Soil Water Conserv. 18(4): 354-363.

King, K.F.S. 1969. Agri-silviculture. The Taungya system Bull. No 7. Department of Forestry, University of Ibadan. 109p.

Lasco, R.D., Delfino, R.J.P., Catacutan, D.C., Simelton, E.S. and Wilson, D.M. 2014. Climate risk adaptation by smallholder farmers: the roles of trees and agroforestry. Current Opinion in Environmental Sustainability, 6: 83-88.

Lipper, L., McCarthy, N., Zilberman, D., Asfaw, S. and Branca, G. 2017. Climate smart agriculture: building resilience to climate change, Springer Nature, 630p.

Liu, C., Plaza-Bonilla, D., Coulter, J.A., Kutcher, H.R., Beckie, H.J., Wang, L., Floc'h, J.B., Hamel, C., Siddique, K.H., Li, L. and Gan, Y., 2022. Diversifying crop rotations enhances agroecosystem services and resilience. Advances in Agronomy, 173, pp.299-335.

Morales, D., Rodríguez, P., Dell'Amico, J., Nicolas, E., Torrecillas, A. and Sanchez-Blanco, M.J. 2003. High-temperature preconditioning and thermal shock imposition affects water relations, gas exchange and root hydraulic conductivity in tomato. Biol. Plant.47: 203-208.

Nagabhatla, N., Sahu, S., Gaetaniello, A., Wen, L. and Lee, W. 2015. Understanding impacts of climate variation in varied socio-ecological domains: A prerequisite for climate change adaptation and management. Handbook of Climate Change Adaptation, Springer International Publishing AG: Berlin, Germany, pp. 589-617.

NOAA National Centers for Environmental Information (2024). Annual 2023 Global Climate Report. Available: https://www.ncei.noaa.gov/access/monitoring/monthly-report/global/202313.

Quandt, A., 2020. Contribution of agroforestry trees for climate change adaptation: narratives from smallholder farmers in Isiolo, Kenya. Agroforestry Systems, 94(6), pp.2125-2136.

Rao, C.S., Kareemulla, K., Krishnan, P., Murthy, G.R.K., Ramesh, P., Ananthan, P.S. and Joshi, P.K. 2019. Agro-ecosystem based sustainability indicators for climate resilient agriculture in India: A conceptual framework. Ecological Indicators, 105: 621-633.

Shanker, A., Shanker, C. and Srinivasarao, C. eds., 2018. Climate Resilient Agriculture: Strategies and Perspectives. BoD–Books on Demand.

Tripathi, R. and Bisen, J. P. 2019. Climate Resilient Agricultural Technologies for Future. Training Manual, Model Training Course on Climate Resilient Agricultural Technologies for Future, ICAR-National Rice Research Institute, Cuttack. pp 1-102.

Wilson, M.L. and VanBuren, R., 2022. Leveraging millets for developing climate resilient agriculture. Current Opinion in Biotechnology, 75, p.102683.

Wu, W. and Ma, B., 2015. Integrated nutrient management (INM) for sustaining crop productivity and reducing environmental impact: A review. Science of the Total Environment, 512, pp.415-427.

Ye, Y., Liang, X., Chen, Y., Liu, J., Gu, J., Guo, R. and Li, L. 2013. Alternate wetting and drying irrigation and controlled-release nitrogen fertilizer in late-season rice. Effects on dry matter accumulation, yield, water and nitrogen use. Field Crops Res. 144: 212-224.

Zahra, N., Hafeez, M.B., Wahid, A., Al Masruri, M.H., Ullah, A., Siddique, K.H. and Farooq, M. 2023. Impact of climate change on wheat grain composition and quality. J. Sci. Food Agric.103(6): 2745-2751.

Zhao, X. and Fitzgerald, M. 2013. Climate change: Implications for the yield of edible rice. PLoSone, 8(6): e66218p.

2

Mitigation and Adaptation Strategies for Climate Change and Soil Health

Priya Maria Jacob

Department of Soil Science and Agricultural Chemistry
College of Agriculture, Vellayani, Trivandrum

Abstract

The intricate connection between soil health and climate change emphasizes how urgent it is to implement sustainable practices and look into the ways in which agriculture, soil health, composition, and functionality are affected by climate change. The soil faces transformative challenges as global temperatures rise and weather patterns change, upsetting ecosystems and jeopardizing its equilibrium. Severe weather events, like floods and droughts, worsen and cause erosion, loss of topsoil, reduced fertility, and a shortage of water for microorganisms and plants. When creating plans to mitigate climate change, it becomes critical to comprehend the ability of soil to sequester carbon. In light of climate change, adaptable agricultural methods based on soil science research are crucial to guaranteeing food security. Soil scientists recommend precision agriculture, enhanced irrigation methods, and resilient crop varieties as ways to lessen the effects on agriculture. Governments, farmers, and other stakeholders must work together to implement these practices globally because healthy soils are crucial for creating a resilient and sustainable future.

Keywords: *Soil health, Precision agriculture, Climate change, Sustainable practices, Resilient future*

Introduction

Soil is a silent foundation and a dynamic part of Earth's complex ecosystem. It sustains life, influences plant growth and is involved in various environmental processes. However, the balance of this vital resource is increasingly threatened by the pervasive effects of climate change, manifested by rising temperatures, changing rainfall patterns and changes in atmospheric conditions. As global temperatures rise and weather patterns change, the relationship between soil and climate poses transformative challenges with far-reaching consequences. This research addresses the complexity of this interaction and examines how climate change affects agriculture, soil health, composition and functionality. Soil health, a holistic concept that includes physical, chemical and biological aspects of soils, is closely linked to climate change. A notable consequence of climate change is the increasing frequency and intensity of extreme weather events such as droughts and floods. These events disrupt the delicate balance of soil ecosystems and lead to erosion, loss of topsoil and reduced fertility. Additionally, increased temperatures contribute to soil moisture depletion, reducing water availability for plants and microorganisms and further compromising soil health. A central issue at the interface between soil science and climate change is the role of soil as a carbon sink. Soils store large amounts of carbon in the form of organic material, thereby influencing the environment.

In light of climate change, soil science research informs adaptive agricultural practices that are crucial to maintaining food security. Soil scientists recommend a number of measures to lessen the effects on agriculture, including resilient crop varieties, enhanced irrigation methods, and precision agriculture. Unavoidable aspects of our day are the unquestionable facts of climate change and its effects on the environment. This chapter explores the critical role that soil plays in both influencing and being influenced by changing climatic conditions. It also delves into the complex relationship between soil science and climate change. Developing strategies to improve soils' capacity to support ecosystems and sustain life requires an understanding of their vulnerabilities and adaptive capacities.

Global warming and greenhouse gas emissions

A major threat to the Earth's climate system is global warming, which is being caused by an unparalleled rise in greenhouse gas emissions. As greenhouse gases like carbon dioxide (CO_2), methane (CH_4) and nitrous oxide (N_2O) build up, heat is trapped in the atmosphere and global temperatures rise as a result. From the pre-industrial era, human activities have significantly increased the concentrations of these gases, according to the Intergovernmental Panel on Climate Change (IPCC) (2021). These activities include the burning of fossil fuels, deforestation, and industrial processes. Unmistakable evidence of human influence on the climate is highlighted in the IPCC's Sixth Assessment Report, which also emphasizes how urgent it is to reduce greenhouse gas emissions in order to prevent future temperature increases and the related climate impacts. To mitigate the negative effects of global warming, efforts must be made to reduce greenhouse gas emissions. One historic global attempt to combat climate change is the Paris Agreement, which was ratified by the United Nations Framework Convention on Climate Change (UNFCCC) in 2015. The UNFCCC (2015) established a global target of 1.5 degrees Celsius, with the aspirational goal of keeping the increase in average global temperature well below 2 degrees Celsius above pre-industrial levels. Reducing emissions, switching to sustainable energy sources, and putting in place global climate policies that work is all necessary to meet these goals.

Carbon dynamics and soil as a carbon sink

According to the IPCC (2018), carbon sinks are reservoirs that absorb and store more carbon than they release. Understanding how soil functions as a source and sink of carbon, playing a critical role in global carbon cycles, requires an understanding of carbon sequestration and its release in the soil. In order to create strategies that effectively reduce the effects of global warming, it is imperative to comprehend the relationship between soil's role as a carbon sink and climate change. The ability of soils to absorb and store carbon, mostly in the form of organic matter, is demonstrated by their role as carbon sinks (Lal, 2004). During photosynthesis, plants take in CO_2 and transform it into organic compounds. Plant roots and residues break down and add to the soil's organic carbon (SOC) content. Anthropogenic carbon emissions can be lessened by well-managed soils, which have the capacity to store large amounts of carbon. Rising temperatures, changed precipitation patterns, and extreme weather events can all have an impact on soil microbial activity, organic matter decomposition rates, and nutrient cycling, despite the fact that soil serves as an essential carbon sink (Wu *et al.,* 2024). The stability of soil carbon stocks may be jeopardized by these modifications, which may also cause stored carbon to leak into the atmosphere. We can improve soil carbon sequestration by using sustainable land management techniques including the application of biochar, afforestation, reforestation, agroforestry, and conservation agriculture (Le Quéré *et al.*, 2018). In addition to storing carbon, these methods enhance soil fertility, structure, and water-holding capacity. Reducing atmospheric CO_2 concentrations through efficient carbon sequestration in soils helps to mitigate climate change.

These practices not only sequester carbon but also improve soil structure, fertility and water retention. Effective carbon sequestration in soils contributes to the reduction of atmospheric CO_2 concentrations, playing a role in mitigating climate change.

Soil health and climate change

The sustainability of an ecosystem depends on the health of the soil, which supports a variety of ecological processes necessary for life as we know it. The Food and Agriculture Organization (FAO) has highlighted the importance of healthy soils for biodiversity conservation, food security, and mitigating the effects of climate change (FAO, 2015). A vital component of the Earth's ecosystem, soil health is essential to maintaining life and a variety of ecological processes. As society bears the heavy costs of environmental degradation, more attention is being paid to the complex relationship between soil health and climate change. In order to effectively mitigate the negative effects of climate change, it is imperative to adopt sustainable soil management practices, which are based on a symbiotic relationship between soil health and climate change.

Because soil contains large amounts of organic matter, it acts as a significant carbon sink (Smith *et al.,* 2007)). Long-term carbon sequestration in the soil is achieved by the breakdown of organic materials by a diverse microbial community in healthy soils (Lal, 2004). Unfortunately, soil degradation is a result of human activities that reduce its ability to store carbon, such as deforestation, intensive agriculture, and inappropriate land use. Amplification of the greenhouse gas effect occurs when this degradation releases carbon dioxide (CO_2) into the atmosphere (Batjes, 1996). In contrast, altered precipitation and temperature patterns are two ways that climate change affects the health of soil. Davidson and Janssens (2006) state that increasing temperatures have an impact on soil microbial activity, nutrient cycling, and the availability of vital minerals. A decrease in soil fertility, increased susceptibility to extreme weather events, and soil erosion can result from changes in precipitation patterns (IPCC, 2019). The intricate interaction between these feedback loops causes problems related to climate change to worsen at the same time that soil health deteriorates.

A multifaceted strategy involving policy changes, sustainable land management practices, and increased public awareness is required to effectively address this complex relationship. Policies that promote afforestation, encourage sustainable agriculture, and aid in conservation efforts must be put in place by governments and international organizations (FAO, 2015). Education is essential for creating a global awareness of the importance of soil health and how it relates to climate change, as well as for motivating people to adopt sustainable lifestyle choices (NRC, 2010). Furthermore, soil science research and technological developments are essential for creating creative ways to improve soil resilience in the face of changing climatic conditions (Lehmann *et al.,* 2015).

Navigating the impacts: Rising temperatures, soil health and agricultural resilience

A prominent and immediate effect of climate change is the gradual increase in global temperatures. Wide-ranging effects of surface warming are being felt by many ecosystems, with soil health being one particularly important element that is temperature-sensitive.

1. ***Impact on soil microbial communities:*** Rising temperatures can significantly affect the composition and activity of soil microbial communities. Soil fauna, which make up 23% of known animal species, play a crucial role in ecosystem functions such as nutrient cycling, microbial regulation and the formation of soil structure. However, global climate change is affecting the density, composition and distribution of soil fauna and flora (Kudureti *et al.*, 2023). Elevated temperatures can accelerate microbial activity and lead to increased decomposition of organic matter (Bardgett *et al.*, 2008). While this initially releases nutrients, prolonged warming could deplete soil organic carbon and affect soil structure and fertility (Fierer *et al.*, 2012).

2. ***Changes in soil moisture dynamics:*** Temperature increases contribute to changes in precipitation patterns and evaporation rates and affect soil moisture dynamics. Higher temperatures can increase evaporation and lead to drier soils in certain regions (Dai, 2013). Combined with changing rainfall patterns, this can lead to increased droughts and fluctuations in soil moisture. Such conditions pose a challenge to plant growth and affect agriculture and ecosystem stability (Seneviratne *et al.*, 2010).

3. ***Soil structure and erosion:*** Higher temperatures can affect soil structure, especially in clay-rich soils. Soil drying and cracking due to increased evaporation can result in reduced water infiltration and increased surface runoff. This, in turn, contributes to soil erosion, reduces nutrient-rich topsoil and impairs the soil's ability to support plant life. Erosion exacerbates the challenges that rising temperatures pose to soil health (Montgomery, 2007).

4. ***Impacts on crop productivity:*** The connection between rising temperatures and soil health is particularly important in agriculture. Increased temperatures can affect plant growth and development, leading to changes in phenology, reduced yields and altered plant distribution patterns (Lobell *et al.*, 2011). Some crops may face challenges adapting to the new thermal conditions, which could have implications for global food production and food security.

5. ***Pests and diseases:*** Climate change is altering the geographical distribution and prevalence of pests and diseases, affecting crop health and productivity (Chakraborty *et al.*, 2018). Warmer temperatures can create favourable conditions for pest spread, leading to heavier infestations and the need for more aggressive pest control strategies. Changing disease patterns may also threaten farmers' livelihoods and further complicate efforts to maintain stable crop production.

6. ***Impact on livelihoods and rural communities:*** The social and economic implications of climate change on agriculture are significant, as highlighted by the IPCC in 2014. Rural communities, which heavily rely on agriculture, are encountering heightened risks of poverty and food insecurity. Among these communities, smallholder farmers are particularly vulnerable due to their limited resources to effectively adapt to changing conditions. To promote resilience and maintain sustainable livelihoods, it is crucial to implement climate-resilient agricultural practices and provide support for rural communities.

Maximizing role of soil in climate change adaptation

Strategies for adaptation are crucial as the effects of climate change intensify to lessen the difficulties that communities, agriculture, and ecosystems around the world face. The soil is a frequently disregarded but essential component of climate change adaptation. This essay explores the diverse functions of soil in adapting to climate change, highlighting its ability to promote resilience, sustainable practices, and the general health of ecosystems.

1. ***Water regulation and drought resilience:*** Soil acts as a natural water regulator and plays a crucial role in controlling water availability (FAO, 2015). Well-structured soils with good organic matter content can effectively capture and store water during periods of rainfall, providing a buffer against drought conditions. This water retention capacity increases the resilience of ecosystems, supports plant life during dry periods and thus contributes to general adaptation to climate change.

2. ***Carbon sequestration for mitigating climate impacts:*** Carbon dioxide from the atmosphere is sequestered by soil, making it a significant carbon sink (Batjes, 1996). This vital role lowers greenhouse gas concentrations, which lessens the effects of climate change. A more stable global climate is promoted and carbon sequestration is improved by sustainable soil management techniques like conservation agriculture, afforestation, and replanting.

3. ***Enhanced nutrient cycling for agricultural resilience:*** Healthy soils support nutrient cycling, a process critical to maintaining plant growth (Lal, 2004). Improved soil health facilitates nutrient availability and reduces reliance on synthetic fertilizers. Adaptive agricultural practices, including crop rotation, cover cropping and organic farming, promote soil health and resilience and make agriculture more adaptable to changing climate conditions.

4. ***Biodiversity support and ecosystem resilience:*** According to Bardgett (2010), soil serves as a thriving and varied environment for microorganisms, fungi, and diverse animal species. This wide array of biodiversity plays a crucial role in maintaining the balance of ecosystems by promoting nutrient cycling, preventing diseases, and ensuring overall ecological stability. By implementing sustainable land management practices that prioritize the preservation of soil biodiversity, we can bolster the resilience of entire ecosystems, enabling them to better withstand the challenges posed by a changing climate.

5. ***Reduced vulnerability to extreme weather events:*** Reducing the risk of flooding and soil erosion, well-structured soils with a high content of organic matter are better able to absorb and hold onto excess water during periods of heavy rainfall (Six *et al.,* 2004). Furthermore, these soils are resistant to extremely high or low temperatures, which helps to stabilize plant roots and increase ecosystem resilience against extreme weather events linked to climate change.

6. ***Supporting sustainable land management practices:*** Soil adaptability is closely linked to sustainable land management practices (Hobbs *et al.*, 2008). Conservation agriculture, agroforestry and measures to prevent soil erosion help maintain soil health and improve its adaptability. These practices promote sustainable land use and support both agricultural productivity and ecosystem services.

7. ***Local climate regulation and microclimate creation:*** Vegetation interacts with the soil and creates microclimates that regulate local temperatures (Akbari *et al.*, 2001). Urban areas with green spaces and well-maintained soils experience cooler temperatures, relieving urban heat islands. These microclimates contribute to climate adaptation efforts by mitigating the effects of extreme heat events.

Sustainable soil management practices

The importance of sustainable soil management techniques has gained attention as the globe struggles to address the effects of climate change. In addition to supporting biodiversity and agriculture, healthy soils are a major source of carbon sequestration and are essential for reducing the effects of climate change. The following are the sustainable soil management techniques that support both adaptation and mitigation of climate change.

1. ***Conservation agriculture:*** Conservation agriculture focuses on reducing soil disturbance by using no-till or reduced tillage techniques. Conservation agriculture prevents soil erosion, improves water retention, and encourages the build-up of organic matter by leaving crop residues on the field surface (Hobbs *et al.*, 2008). By enhancing general soil health and aiding in the sequestration of soil carbon, these techniques strengthen agriculture's resistance to climate change.
2. ***Biochar application:*** Biochar shows significant potential for both emissions reduction and carbon dioxide (CO_2) removal, offering a promising route to addressing climate challenges. Globally, biochar systems have the potential to achieve emissions reductions of 3.4–6.3 PgCO_2e, with half of this representing actual CO_2 removal. The longer persistence of biochar compared to the biomass from which it is derived plays a crucial role in achieving significant CO_2 removal (Lehmann *et al.,* 2021).

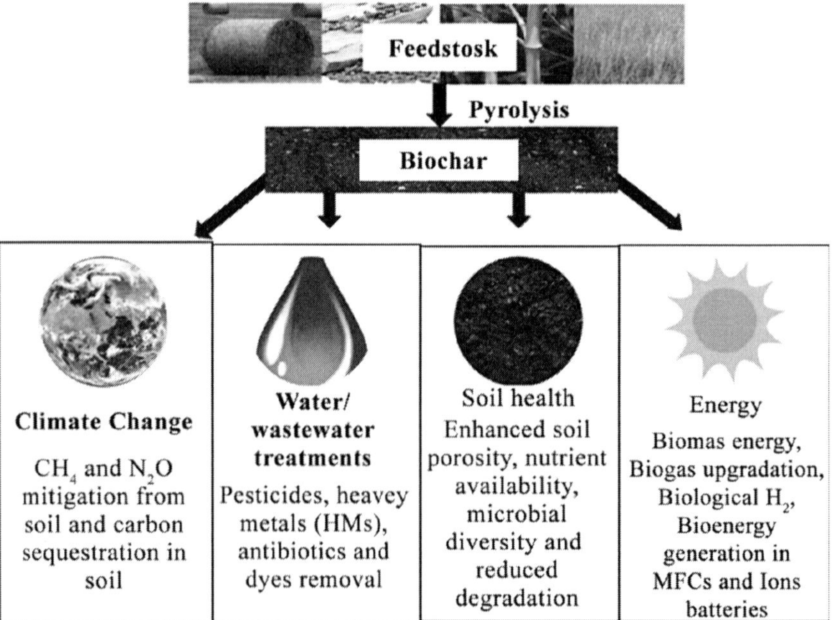

Fig. Biochar and Environmental Sustainability (Malyan *et al.,* 2021)

3. ***Cover Cropping:*** Intercropping involves growing crops specifically to cover and protect the soil between the main growing seasons. Cover crops prevent erosion, suppress weed growth, and add organic matter to the soil when incorporated. Additionally, certain cover crops such as legumes help nitrogen fixation, reducing the need for synthetic fertilizers. These practices improve soil fertility, improve water infiltration, and increase resilience to climate-related challenges.

3. ***Crop rotation:*** Crop rotation is a traditional agricultural practice in which different crops are grown in a specific order on the same piece of land. This helps break pest and disease cycles, improves soil structure and promotes nutrient cycling (jose, 2003). Diverse crop rotations contribute to overall soil health, reduce dependence on chemical inputs and make agricultural systems more adaptable to changing climate conditions.

4. ***Agroforestry:*** Agroforestry integrates trees or shrubs into agricultural landscapes and offers numerous benefits for the soil and climate. Trees help sequester carbon, reduce soil erosion and improve water retention. Agroforestry systems can improve the microclimate and make it more resilient to extreme weather events (Jose, 2009). Combining woody perennials with crops or livestock promotes biodiversity and sustainable land use.

5. ***Organic farming:*** Organic farming focuses on the health of the soil by avoiding synthetic chemicals and promoting natural processes. By avoiding the use of synthetic fertilizers and pesticides, organic farming promotes microbial diversity in the soil and improves nutrient cycling (Reganold *et al.,* 2010). Adding organic matter through composting and manure helps sequester carbon in the soil and makes organic systems more resilient to climate change.

6. ***Nutrient management:*** For sustainable soil management, nutrient utilization must be optimized. By minimizing excess fertilizer use, nutrient runoff and the related environmental effects, precision farming techniques—like site-specific nutrient applications—help protect the environment (Yadav *et al.,* 2023). Efficient nutrient utilization, enhanced soil health, and reduced environmental degradation are ensured when nutrient inputs and crop requirements are balanced.

7. ***Mulching:*** Mulching is an agricultural and gardening practice that involves covering the soil with a layer of organic or inorganic material. This practice helps conserve soil moisture by reducing evaporation, preventing soil erosion and maintaining more stable soil temperatures.

8. ***Drought-resistant Crops:*** Planting crop varieties that are adapted to water-scarce conditions enhances soil resilience during drought periods.

   Drought-Tolerant Maize (DTM): DTM cultivars have been developed to endure water scarcity and function optimally in environments with less water. These cultivars can flourish in arid or semi-arid areas because they frequently have characteristics like deep root systems and efficient water use (CIMMYT, 2020).

   Drought-Resistant Wheat varieties: Researchers have created drought-resistant wheat types that can continue to produce even in situations where there is a shortage of water. Wheat is a basic crop. Improved water efficiency and resistance to water stress throughout crucial growth phases are among the characteristics these cultivars frequently have. Examples are cultivars bred for increased drought resilience and water efficiency, such as "Drysdale" and "Seriaph" (Reynolds and Tuberosa, 2008).

   Drought-Tolerant Rice cultivars: Cultivars of rice that are resistant to drought have been developed in areas where the crop is a major source of water scarcity. These cultivars frequently exhibit traits including lowered water requirements, modified flowering schedules, and improved resilience to stress brought on by drought. 'Swarna-Sub1' is one of the varieties that has been created to resist drought and submersion (Neeraja *et al.*, 2007).

   Drought-Adapted Legumes (e.g., Chickpeas and Lentils): Legume crops, such as lentils and chickpeas, have been engineered to withstand water scarcity better. These types frequently have characteristics like deep roots that enable them to reach water from deeper soil levels and water-saving devices. Lentil variants like 'CDC Robin' and chickpea cultivars like 'ICC 4958' are bred to have better resistance to drought (Varshney *et al.*, 2014).

9. ***Water conservation practices:*** Water conservation measures play a critical role in addressing the challenges of climate change, particularly related to changing rainfall patterns, the increasing frequency of droughts and general water scarcity. These practices contribute to climate change adaptation and mitigation. Implementing efficient irrigation methods such as drip irrigation or rainwater harvesting, water recycling and smart water management technologies help optimize water use and maintain soil moisture levels.

Conclusion

The necessity of sustainable practices is highlighted by the connection between soil health and climate change. We can create comprehensive plans for both resilient ecosystems and climate change by comprehending their relationship. For a sustainable future, cooperation, educated policy, and conscientious land management are essential. The complex interplay among soil, carbon dynamics, and climate change emphasises how crucial sustainable land management is. Preserving the ability of soil to sequester carbon contributes to the general resilience and health of the soil. For the purpose of solving climate change concerns, a holistic approach is essential. Growing global temperatures have an adverse effect on crop yield, moisture content, and microbial activity in the soil. It is essential to give sustainable land management first priority. For resilient ecosystems, understanding the symbiotic link between soil health and temperatures is essential. Climate change presents interrelated concerns related to water, drought, and soil resilience. Resilience requires putting adaptive measures into practice, encouraging sustainable water management, and placing a high priority on soil conservation. It is critical to understand how water, drought, and soil health interact. Soil is a resource that is resilient to climate change. It is essential for controlling water, storing carbon, and preserving biodiversity. A more resilient future depends on adaptation measures that acknowledge the significance of soil. Food security and agriculture are threatened by climate change. For a resilient and food-secure future, immediate action is required, including global cooperation, technology, and sustainable behaviours. Managing soil sustainably is essential to combating climate change. Resilience building and improved soil health provide a comprehensive strategy for sustainable agriculture. To put these practices into action for a resilient and sustainable future, global cooperation is required.

References

Akbari, H., Pomerantz, M., & Taha, H. (2001). Cool surfaces and shade trees to reduce energy use and improve air quality in urban areas. Solar Energy, 70(3), 295-310.

Bardgett, R.. (2010). The Biology of Soil: A Community and Ecosystem Approach. The Biology of Soil: A community and ecosystem approach. 1-256. 10.1093/acprof:oso/9780198525035.001.0001.

Bardgett, R. D., van der Putten, W. H., & Below-ground biodiversity and ecosystem functioning. (2008). Nature, 515(7528), 505-511.

Batjes, N. H. (1996). Total carbon and nitrogen in the soils of the world. European Journal of Soil Science, 47(2), 151-163.

Chakraborty, S., Tiedemann, A. V., & Teng, P. S. (2000). Climate change: potential impact on plant diseases. Environmental Pollution, 108(3), 317-326.

CIMMYT. (2020). Drought-Tolerant Maize for Farmers Facing Climate Uncertainty. Retrieved from https://www.cimmyt.org/what-we-do/drought-tolerant-maize-for-farmers-facing-climate-uncertainty/)

Dai, A. (2013). Increasing drought under global warming in observations and models. Nature Climate Change, 3(1), 52-58.

Davidson, E. A., & Janssens, I. A. (2006). Temperature sensitivity of soil carbon decomposition and feedbacks to climate change. Nature, 440(7081), 165-173.

FAO. (2015). Global Soil Partnership. Retrieved from http://www.fao.org/global-soil-partnership/en/

Fierer, N., Bradford, M. A., & Jackson, R. B. (2012). Toward an ecological classification of soil bacteria. Ecology, 88(6), 1354-1364.

Haowei Wu, Huiling Cui, Chenxi Fu, Ran Li, Fengyuan Qi, Zhelun Liu, Guang Yang, Keqing Xiao, Min Qiao. (2024). Unveiling the crucial role of soil microorganisms in carbon cycling: A review. Science of The Total Environment, 909(168627).

Hobbs, P. R., Sayre, K., & Gupta, R. (2008). The role of conservation agriculture in sustainable agriculture. Philosophical Transactions of the Royal Society B: Biological Sciences, 363(1491), 543-555.

IPCC. (2014). Climate Change 2014: Impacts, Adaptation and Vulnerability. Part A: Global and Sectoral Aspects. Contribution of Working Group II to the Fifth Assessment Report of the Intergovernmental Panel on Climate Change. Cambridge, United Kingdom and New York, NY, USA: Cambridge University Press.

IPCC. (2018). Global Warming of 1.5°C. Special Report. Retrieved from https://www.ipcc.ch/sr15/

IPCC. (2019). Climate Change and Land: An IPCC special report on climate change, desertification, land degradation, sustainable land management, food security and greenhouse gas fluxes in terrestrial ecosystems. Retrieved from https://www.ipcc.ch/srccl/

IPCC. (2021). Climate Change 2021: The Physical Science Basis. Contribution of Working Group I to the Sixth Assessment Report of the Intergovernmental Panel on Climate Change. Cambridge University Press.

Jose, S. (2009). Agroforestry for ecosystem services and environmental benefits: an overview. Agroforestry Systems, 76(1), 1-10.

Kudureti, A., Zhao, S., Zhakyp, D. et al. Responses of soil fauna community under changing environmental conditions. J. Arid Land 15, 620–636 (2023). https://doi.org/10.1007/s40333-023-0009-4.

Lal, R. (2004). Soil carbon sequestration to mitigate climate change. Geoderma, 123(1–2), 1–22.

Lal, R. (2005). Soil erosion and carbon dynamics. Soil and Tillage Research, 81(2), 137-142. Top of Form

Lehmann, J., Cowie, A., Masiello, C.A. et al. Biochar in climate change mitigation. Nat. Geosci. 14, 883–892 (2021). https://doi.org/10.1038/s41561-021-00852-8

Lehmann, J., Gaunt, J., & Rondon, M. (2015). Bio-char sequestration in terrestrial ecosystems: a review. Mitigation and Adaptation Strategies for Global Change, 11(2), 395-419.

Le Quéré, C., Andrew, R. M., Friedlingstein, P., Sitch, S., Pongratz, J., Manning, A. C., ... & Canadell, J. G. (2018). Global carbon budget 2017. Earth System Science Data, 10(1), 405–448.

Lobell, D. B., Schlenker, W., & Costa-Roberts, J. (2011). Climate trends and global crop production since 1980. Science, 333(6042), 616-620.

Malyan, S.K., Kumar, S.S., Fagodiya, R.K., Ghosh, P., Kumar, A., Singh, R. and Singh, L., 2021. Biochar for environmental sustainability in the energy-water-agroecosystem nexus. Renewable and Sustainable Energy Reviews, 149, p.111379.

Montgomery, D. R. (2007). Soil erosion and agricultural sustainability. Proceedings of the National Academy of Sciences, 104(33), 13268-13272.

Neeraja, C. N., Maghirang-Rodriguez, R., Pamplona, A. M., Heuer, S., Collard, B. C. Y., Septiningsih, E. M., Vergara, G., & Sanchez, D. (2007). A Marker-Assisted Backcross Approach for Developing Submergence-Tolerant Rice Cultivars. Theoretical and Applied Genetics, 115(6), 767–776. https://doi.org/10.1007/s00122-007-0602-7).

NRC (National Research Council). (2010). Advancing the Science of Climate Change. National Academies Press. Retrieved from https://www.nap.edu/read/12782/chapter/1

Reganold, J. P., Elliott, L. F., & Unger, Y. L. (1987). Long-term effects of organic and conventional farming on soil erosion. Nature, 330(6146), 370-372.

Reynolds, M. P., & Tuberosa, R. (2008). Translational Research Impacts on Wheat Productivity in Developing Countries. Journal of Experimental Botany, 60(6), 1533–1543. https://doi.org/10.1093/jxb/ern093)

Seneviratne, S. I., Corti, T., Davin, E. L., Hirschi, M., Jaeger, E. B., & Lehner, I. (2010). Investigating soil moisture–climate interactions in a changing climate: A review. Earth-Science Reviews, 99(3-4), 125-161.

Six, J., Frey, S. D., Thiet, R. K., & Batten, K. M. (2006). Bacterial and fungal contributions to carbon sequestration in agroecosystems. Soil Science Society of America Journal, 70(2), 555-569.

Smith, P., D. Martino, Z. Cai, D. Gwary, H. Janzen, P. Kumar, B. McCarl, S. Ogle, F. O'Mara, C. Rice, B. Scholes, O. Sirotenko, 2007: Agriculture. In Climate Change 2007: Mitigation. Contribution of Working Group III to the Fourth Assessment Report of the Intergovernmental Panel on Climate Change [B. Metz, O.R. Davidson, P.R. Bosch, R. Dave, L.A. Meyer (eds)], Cambridge University Press, Cambridge, United Kingdom and New York, NY, USA.

Varshney, R. K., Thudi, M., Nayak, S. N., Gaur, P. M., Kashiwagi, J., Krishnamurthy, L., & Jaganathan, D. (2014). Genetic Dissection of Drought Tolerance in Chickpea (Cicer arietinum L.). Theoretical and Applied Genetics, 127(2), 445–462. https://doi.org/10.1007/s00122-013-2230-6.

Yadav, A., Yadav, K., Ahmad, R., & A., K. (2023). Emerging Frontiers in Nanotechnology for Precision Agriculture: Advancements, Hurdles and Prospects. Agrochemicals, 2(2), 220-256. https://doi.org/10.3390/agrochemicals2020016.

3

Mitigation and Adaptation Strategies for Climate Change and Organic Agriculture

Muthyala Abhiram, Krishnapriya M K, Naveen Leno and Dhanesh Kumar T.V.

Department of Soil Science and Agricultural Chemistry
College of Agriculture, Vellayani, Kerala Agricultural University

Abstract

Climate change, characterized by long-term shifts in temperature and weather patterns, profoundly affects soil health, making it a critical consideration for agriculture, biodiversity, carbon cycling, and ecosystem resilience. This chapter explores the evidences and effects of climate change, emphasizing rising greenhouse gas concentrations and their impact on regions worldwide. Soil health, the enduring capability of soil to function as a thriving ecosystem, is assessed through measurable physical, chemical, and biological attributes. Climate-induced alterations, such as heavy precipitation, altered monsoon rainfall, rising temperatures, and forest fires, significantly impact soil physical, chemical, and biological properties. These changes affect topsoil aggregates, soil erosion, nutrient availability, and microbial communities. Moreover, the chapter introduces the concept of climate smart soils, emphasizing mitigation of greenhouse gas emissions through practices promoting soil health. Technologies like bioenergy carbon capture and storage, conservation agriculture, and organic amendments are essential components of climate smart soils. The importance of organic agriculture in mitigating climate change is highlighted, offering a sustainable approach to preserving soil health. The chapter concludes by advocating for future research to explore predictive modelling and adaptive strategies, ensuring informed practices to mitigate climate change impacts on soil ecosystems. Overall, maintaining soil health is critical for sustainable climate change mitigation and adaptation.

Introduction

Climate plays a crucial role in shaping the agricultural landscape of a region, and the changing climatic conditions pose a significant threat to global food security. In developing countries in Asia and Africa, the primary challenges to realizing optimal crop yields are declining soil fertility, particularly soil organic carbon and drought. The unchecked emission of greenhouse gases is expected to lead to a considerable rise in the average global temperature by 6°C in the next century, primarily driven by activities such as fossil burning and urbanization. Over the past centuries, there has been a noticeable increase in the concentration of CO_2 and CH_4, with up to a 35% expected rise in nitrous oxide due to the improper use of nitrogenous fertilizers. Climate models predict a substantial impact on agricultural productivity in the 21st century, with rising temperatures and more frequent extreme weather events (NASA, 2022). To maintain agricultural productivity, it is imperative to understand the effects of increasing temperatures, shifting precipitation patterns, and rising CO_2 levels on soil health. It is noteworthy that climate is one of the key factors influencing soil formation, alongside four other factors. Temperature and precipitation directly affect soil formation by providing the necessary conditions for weathering and biomass production. The sum of active temperature and precipitation-evaporation ratio are critical parameters that determine energy consumption for soil formation, water balances in soil, organic-mineral interactions, and the transformation processes within the soil. However, human activities such as population growth, deforestation, disruption of marine ecosystems, and the greenhouse effect contribute to climate change, leading to denudation, depletion of soil organic matter, destruction of soil structure, erosion, and desertification, ultimately causing a decline in soil health and environmental degradation (IMD, 2021). Despite being a gradual process, climate change, characterized by small, long-term changes in temperature and precipitation, significantly influences various soil processes, particularly those related to soil fertility. The anticipated effects of climate change on soils primarily involve alterations in soil moisture conditions, increased soil temperature, and elevated CO_2 levels. Understanding the consequences of climate change on soil health is crucial, as it has far-reaching impacts on agriculture, biodiversity, carbon cycling, and ecosystem resilience in the face of an evolving climate (Nunez *et al.,* 2019).

1. Climate change

Climate change pertains to prolonged alterations in temperatures and atmospheric conditions. It is a worldwide occurrence that has been ongoing since the inception of Earth. While the planet has experienced distinct cold and warm cycles throughout its climate history, the shifts have been notably accelerated over the last 150-200 years on a global scale. (Fauchereau *et al.,* 2003). Climate change is mainly caused by the 'greenhouse effect'. The Sun emits energy in the form of sunlight, primarily in the visible and ultraviolet spectrum. This solar radiation reaches the Earth's atmosphere. A portion of this incoming solar radiation is absorbed by the Earth's surface, warming it. Some of the energy is then radiated back into space as heat, mainly in the form of infrared radiation. The Earth's surface emits heat in the form of infrared radiation. Greenhouse gases in the atmosphere, such as carbon dioxide (CO_2), methane (CH_4), and water vapour (H_2O), absorb and re-radiate some of this heat energy back toward the Earth's surface. This process creates a blanket effect, trapping heat within the Earth's atmosphere. As a result, the Earth's surface temperature is higher than it would be if these greenhouse gases were not present. This natural greenhouse effect is crucial for maintaining Earth's temperature at a level suitable for life. Without greenhouse effect, the earth's temperature would be around -18^{0}C. More greenhouse gases are released into the atmosphere due to anthropogenic activities resulting in global warming (higher temperatures, increased warming and damages balance) leading to climate change (Vijayavenkataraman *et al.,* 2012).

Global warming and greenhouse gases

Human activities have been identified as significant contributors to climate change over the past 200 years (Alirezaei *et al.,* 2017). These activities have led to alterations in the chemical composition of the atmosphere. Greenhouse gases (GHGs) and the Earth's magnetic field are identified as the primary factors influencing global climate change (Chen *et al.,* 2016). The increasing emissions of greenhouse gases, coupled with changes in the Earth's magnetic field, have resulted in a reduced reflection of radiation from the Earth's surface. To restore the balance between incoming and outgoing radiation and counteract this anomaly, the Earth's climate undergoes changes. The growing emissions of greenhouse gases lead to more radiation being trapped within the Earth's surface. These gases, including carbon dioxide, methane, nitrous oxide, water vapor, and hydrofluorocarbons, contribute to the warming of the Earth's surface by permitting sunlight to enter while impeding the escape of heat, resulting in global warming with far-reaching consequences (Huang *et al.,* 2016).

Numerous researchers and scientists globally have conducted experiments and analyses to identify ways to reduce greenhouse gas emissions and mitigate the impacts of global warming, aiming to maintain the delicate balance of life on Earth (Mikhaylov *et al.,* 2020). Greenhouse gases, both natural and synthetic, play a crucial role in trapping heat in the atmosphere and contributing to planetary warming. Carbon dioxide is absorbed by various carbon sinks like forests, soil, and the ocean, while fluorinated gases are eliminated by sunlight in the upper atmosphere. The specific chemical properties of different greenhouse gases influence their removal processes from the atmosphere. Greenhouse gases, characterized by a molecular structure with three or more atoms, create a continuous cycle of trapping heat and contributing to a global temperature increase, akin to the functioning of a greenhouse (Archer, 2007; Chultheis, 2013).

The Global Warming Potential (GWP) is a comprehensive measure of a gas's impact on global warming, considering its ability to trap solar radiation as heat and its atmospheric persistence. The GWP is typically calculated in relation to carbon dioxide's GWP, providing a standardized measure for assessing the overall warming potential of different gases.

Table 1: Variability in the status of emission of greenhouse gases

Greenhouse gas	Preindustrial (1000-1750)	1998	2023	Atmospheric lifetime (Years)	100 Year Global Warming Potential (GWP)
Carbondioxide (CO_2)	280 ppm	365 ppm	421.71 ppm	50-200	1
Methane (CH_4)	0.7 ppb	1745 ppb	1923.6 ppb	12	23
Nitrous oxide (N_2O)	270 ppb	314 ppb	335.2 ppb	114	296
Sulfurhexafluoride (SF_6)	0	4.2ppt	10.85 ppt	3200	22,200

ppm – parts per million, ppb – parts per billion, ppt – parts per trillion

Source: (NASA, 2022)

The primary sources of greenhouse gas emissions by economic sector in India (IPCC 2021): Electricity production contributes to the second-largest share of greenhouse gas emissions, accounting for 44 percent. This includes emissions from the generation of electric power used by various sectors, such as industry. Manufacturing industries and construction contribute 18 percent of greenhouse gas emissions, primarily resulting from the combustion of fossil fuels for energy and certain chemical reactions involved in transforming raw materials into goods. Transportation is responsible for 13 percent of greenhouse gas emissions, mainly due to the burning of fossil fuels for cars, trucks, ships, trains, and planes. More than 94 percent of the fuel used in transportation is petroleum-based, predominantly gasoline and diesel. Agriculture contributes

14 percent of greenhouse gas emissions, originating from sources like livestock, agricultural soils, and rice production. The remaining 16 percent is distributed between Industrial processes and product use (13 percent) and Waste (3 percent).

2. Evidences for climate change

As per the findings of the Intergovernmental Panel on Climate Change (IPCC), the impact of human activities on the warming of the climate system has transitioned from being a theoretical concept to an established fact since the initiation of systematic scientific assessments in the 1970's. Information derived from diverse natural sources, including ice cores, rocks, and tree rings, along with data collected through modern technologies such as satellites and instruments, consistently demonstrates indicators of a climate in flux. Whether it's the global increase in temperatures or the melting of ice sheets, there is ample evidence pointing towards a planet undergoing warming.

Raised atmospheric CO_2

Based on the comparison of atmospheric samples contained in ice cores and more recent direct measurements, provides evidence that atmospheric CO_2 has increased to 420 parts per million since the industrial revolution. The current warming trend is different because it is clearly the result of human activities since the mid-1800s, and is proceeding at a rate not seen over many recent millennia. It is undeniable that human activities have produced the atmospheric gases that have trapped more of the sun's energy in the earth system. This extra energy has warmed the atmosphere, ocean, and land, and widespread and rapid changes in the atmosphere, ocean, cryosphere, and biosphere have occurred (Jiang and Khan, 2023)

Rise in global temperature

The planet's average surface temperature has risen by about 1.1°C since preindustrial times, a change driven largely by increased carbon dioxide emissions into the atmosphere and other human activities. Most of the warming occurred in the past forty years, with the seven most recent years being the warmest. The years 2016 and 2020 are tied for the warmest year on record (Yashmin *et al.,* 2022).

Raised atmospheric methane

Methane (CH_4) is a potent greenhouse gas, ranking as the second-largest contributor to climate warming after carbon dioxide (CO_2). While a methane molecule is more effective at trapping heat compared to a CO_2 molecule, it has a relatively short atmospheric lifespan of 7 to 12 years, whereas CO_2 can persist for hundreds of years. Methane emissions originate from both natural sources and human activities, with an estimated 60 percent of current emissions attributed to human actions. The atmospheric concentration of CH_4 reached 1923.6 ppm by 2022. The primary contributors to methane emissions are agriculture, fossil fuel activities, and the decomposition of landfill waste. Paddy fields and livestock production account for 17 percent and 55 percent, respectively, of global CH_4 emissions. Natural processes contribute 40 percent to methane emissions, with wetlands being the largest natural source (Bai *et al.,* 2020).

Ocean warming

Ninety percent of the global warming is taking place in the ocean, leading to an increase in the internal heat of the water. The rates of ocean warming are measured in zettajoules, and since 1955, ocean warming has risen to 345 zettajoules. The heat stored in the ocean results in the expansion of water, contributing to one-third to one-half of the global rise in sea levels. The majority of this added energy is retained at the surface, within the depth range of zero to 700 meters. The past decade has been identified as the warmest for the ocean since at least the 1800s, with the year 2022 marking the warmest recorded year for the ocean, coinciding with the highest global sea level. Spanning over 70 percent of the Earth's surface, the global ocean possesses a remarkably high heat capacity. It has absorbed 90 percent of the warming attributed to the increasing levels of greenhouse gases in recent decades, with the top few meters of the ocean storing as much heat as the entire Earth's atmosphere. The consequences of ocean warming encompass sea level rise due to thermal expansion, coral bleaching, accelerated melting of major ice sheets, heightened hurricane activity, and alterations in ocean health and biochemistry (Cheng *et al.,* 2022).

Shrinking ice sheets

Antarctica is experiencing ice mass loss, or melting, at an annual average rate of approximately 150 billion tons, while Greenland is losing about 270 billion tons each year, contributing to the overall rise in sea levels. The combined ice sheet shrinkage is estimated at 424 billion metric tons annually. The meltwater originating from these ice sheets accounts for approximately one-third of the global average increase in sea levels since 1993 (Scambos and Moon, 2022).

Glacial retreat

Glacial retreat is the term used to describe the reduction or shrinking of a glacier over time, resulting from either decreased ice accumulation or increased ice melting. While glacial retreat is a natural occurrence, it has accelerated notably in recent decades due to the impacts of climate change. Alaska hosts over 10% of the Earth's glacial area outside of the ice sheets. The maritime glaciers in Alaska have exhibited variable patterns of advancement and retreat since the 1980s (Kleber *et al.,* 2023).

Rise in sea level

The rise in global sea levels, attributed to human-induced global warming, is occurring at rates unprecedented in the last 2,500 years. Since 1993, sea levels have increased by 98 mm. Two primary factors linked to global warming contribute to this phenomenon: the additional water from the melting of ice sheets and glaciers, and the expansion of seawater as it warms (Saintilan *et al.,* 2022).

The effects of climate change

Drought

Drought is a term with varied definitions across different fields. In meteorology, it is characterized as an extended period of minimal or no rainfall, referred to as meteorological drought. From a biological standpoint, the impact of insufficient rainfall (or limited rainfall) on plant life is considered, involving reduced water potential in plant tissues due to a deficit in soil moisture (hydrological drought) resulting from a period of meagre or no rainfall. In agriculture, where the focus is on crop yield, drought is defined as a period of water scarcity leading to soil moisture deficit, ultimately adversely affecting plant yield. The drought-prone area in India has witnessed a 57 percent increase since 1997 (WMO, 2020).

Wild Fire

Wildfires play a crucial role in modifying ecosystems, influencing the biological, chemical, and physical characteristics of forest soils. Fires of high intensity are particularly known for completely burning organic matter, causing severe detrimental effects on forest soils. These intense fires lead to the volatilization of nutrients, the disruption of soil aggregate stability, an elevation in soil bulk density, an increase in soil particle hydrophobicity resulting in reduced water infiltration, heightened erosion, and the destruction of soil biota (Agbeshie *et al.,* 2022). Over the past two decades, from 2000 to 2020, the frequency of forest fire incidents in India has seen a notable increase of 52 percent (Alexakis *et al.,* 2021).

Heavy precipitation

Heavy precipitation occurs when the volume of rain or snow in a particular location significantly surpasses the usual levels. The intensity and frequency of precipitation can be influenced by climate change. Warmer oceans contribute to increased water evaporation into the atmosphere. As moisture-rich air converges over land or forms storm systems, it can result in more pronounced and frequent precipitation. Climate change has the potential to impact both the intensity and occurrence of precipitation. According to Nayak (2023), there is a projected increase in rainfall over India, with a 4-5 percent rise by 2030 and a 6-14 percent increase by 2080 compared to the levels observed during 1961-1990 (1034 mm).

Floods

As per the National Flood Commission, over 40 million hectares, out of the total geographical area of 329 million hectares, are susceptible to flooding. Floods represent a recurring event, resulting in substantial loss of lives and damage to livelihoods, properties, infrastructure, and public utilities. On an annual basis, an average of 18.6 million hectares of land is impacted by floods. The yearly average cropped area affected stands at approximately 3.7 million hectares (Mohanty *et al.,* 2020).

Heat waves

A heat wave is characterized by an unusually high temperature, surpassing the normal maximum for the summer season, particularly in the North-Western regions of India. In the plains, a heat wave is identified when the maximum temperature of a station reaches 40 °C or higher, while in hilly regions, it is considered when the temperature reaches 30 °C or above. Typically occurring between March and June, heat waves can occasionally extend into July (Kiarsi *et al.,* 2023). Prolonged exposure to extreme heat can have various impacts, including crop damage, harm or mortality to livestock, and an elevated risk of wildfires.

Sea water intrusion/ inundation

Seawater intrusion refers to the inward movement of seawater into coastal aquifers, representing a significant factor in the deterioration of coastal groundwater resources. This intrusion not only adversely impacts industrial and agricultural development in the area but also undermines the living standards of the local population (Demirel, 2004). Coastal aquifers are highly vulnerable to both regional and global factors, encompassing sea-level rise, storm surges, changes in climatic conditions, shoreline erosion, and coastal flooding. Among these factors, climate change and rising sea levels emerge as the most crucial climatic elements influencing seawater intrusion in coastal regions (Prusty and Farooq, 2020).

Glacial lake outburst flood

A Glacial Lake Outburst Flood (GLOF) is a sudden and potentially disastrous flood that occurs when water stored behind a glacier or a moraine, a natural accumulation of ice, sand, pebbles, and debris, is released rapidly. These floods occur when glacial lakes, formed by melting ice, accumulate water behind vulnerable moraine dams. Unlike robust earthen dams, these moraine dams can fail abruptly, releasing large volumes of water within minutes to days, resulting in severe downstream flooding (Taylor *et al.,* 2023). The Himalayan terrain, characterized by steep mountains, is particularly susceptible to GLOFs. The process of glacier melting in the Sikkim Himalayas has been accelerated by climate change and rising global temperatures. The region now hosts over 300 glacial lakes, with ten identified as prone to outburst floods. GLOFs can be triggered by various factors, including earthquakes, extremely heavy rainfall, and ice avalanches. These events have the potential to release millions of cubic meters of water in a short timeframe, with peak flows recorded during GLOFs reaching as high as 15,000 cubic meters per second (NDMA, 2021).

Impact of climate change on soil physical properties

Soil structure

During rainfall, topsoil aggregates undergo fragmentation into micro-aggregates and fine particles, impacting soil stability and porosity. The productivity of soil is closely tied to aggregate stability and porosity, influencing overall soil health and fertility. However, intense rainfall can lead to devastating floods, causing soil erosion and landslides. The susceptibility of soil to erosion and landslides depends on factors like aggregate stability and shear strength. Soil erosion is a global challenge, affecting different regions disproportionately. In Kerala, India, abnormal rainfall in 2018 resulted in severe flooding, increasing soil erosion rates significantly. A study by Chinnasamy *et al.* (2020) using the Universal Soil Loss Equation (USLE) revealed an 80% average increase in soil erosion rates during the floods, with the district of Idukki experiencing a staggering 220% increase. The USLE, incorporating factors like rainfall erosivity, soil erodibility, slope steepness, slope length, cover and management, and support practices, helps estimate soil loss. High silt fraction, particularly in Idukki, contributed to increased soil erodibility and disintegration, emphasizing the role of aggregate stability. Moreover, Kerala witnessed a high number of landslides, with continuous rainfall being a key trigger. Landslides in 2019 were attributed to structurally disturbed weakened bedrocks, reduced cohesive strength, and super-saturation of underlying earth material. Factors like steep topography, deep weathering, and structural discontinuities made certain areas more prone to mass movements during heavy rains (Wadhawan *et al.,* 2020). The interplay between soil structure, rainfall, and geological factors underscores the complexity of managing soil stability and preventing erosion-related disasters.

Soil texture

The cohesive forces binding soil particles are predominantly influenced by factors such as particle size distribution, soil organic matter, cation exchange capacity and Fe/Al oxides. Among these, particle size distribution stands out as a key physical indicator affecting cohesive force. Li *et al.* (2023) observed that soils with low clay content, such as sandy loam, silty loam, and loam, tend to develop surface crusts more readily with increased rainfall duration compared to clay loam and silty clay loam. The study ranked soil erodibility based on texture, revealing that sandy loam exhibited the highest erodibility, followed by silty loam, loam, clay loam, and silty clay loam. This underscores the significance of soil texture in understanding and predicting splash erosion rates under varying conditions.

Soil water content

Soil water content (SWC) serves as a vital component of integrative soil health assessments, with its availability responding rapidly to climate variations, especially extreme rainfall or drought events. Managing water requirements for crops necessitates a thorough evaluation of SWC trends over time under the influence of climate change. The impact of climate on SWC is primarily linked to alterations in precipitation and temperature. Zare *et al.* (2022), using the Soil Water Assessment Tool (SWAT) and Regional Climate Models (RCM), projected a future scenario where mean temperatures are expected to rise, accompanied by increased precipitation. Surprisingly, SWC is anticipated to decrease due to a rise in potential evapotranspiration, despite the projected increase in precipitation.

Soil temperature dynamics and environmental influence

Soil temperature, subject to seasonal and daily variations, plays a crucial role in governing physiochemical and biological processes within the soil. It influences gas exchange processes between the atmosphere and the soil, affecting organic matter decomposition, mineralization, soil water content, and plant availability. Different crops require specific soil temperatures for optimal seed germination, with an ideal range of 20-25 °C. Soil surface temperature, when warmer than the air, contributes to heat exchange with the lower atmosphere, intensifying and spreading hot air extremes and heatwaves. Regional differences between air and soil hot extremes have implications for agriculture and terrestrial ecosystem functions (Miralles *et al.,* 2014).

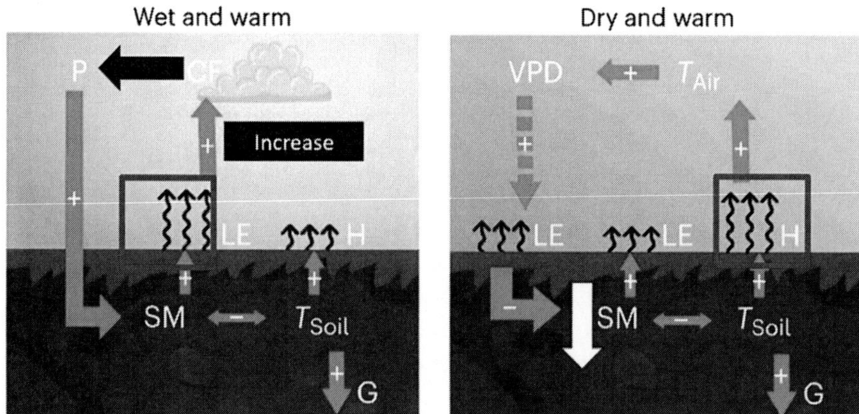

Fig. 1: Soil temperature in the soil moisture-temperature feedback (Miralles *et al.,* 2014)

Soil bulk density and water holding capacity

Soil bulk density, a key indicator routinely assessed in agriculture, reflects soil compactness influenced by land use management. It is intricately linked to soil health indicators such as aeration and infiltration. Bulk density is often negatively correlated with soil organic matter content, and the loss of organic carbon due to increased decomposition from elevated temperatures can lead to increased bulk density. Verma and Jayakumar (2018) investigated the impact of forest fires on soil properties, revealing a reduction in water holding capacity and an increase in bulk density with the frequency of fires. This underscores the susceptibility of soil to compaction under the combined stresses of land management activities and climate change.

Impact of climate change on soil chemical properties

Soil pH

Soil pH is pivotal in the context of climate change, with increased rainfall and elevated carbon dioxide levels leading to soil acidification. Erratic precipitation patterns and rising atmospheric CO_2 contribute to nutrient leaching and reduced pH, impacting plant nutrient availability and soil microbial communities.

Kuttanad, known as the rice bowl of Kerala, experiences acidification attributed to receding seawater influenced by climate change.

Acid sulphate soils in Kuttanad

Kuttanad, below Mean Sea Level, faces unique challenges with soil pH ranging from 3.0 to 4.0, indicative of acid sulphate soils. Climate-induced seawater changes intensify soil acidity, impacting organic carbon content and electrical conductivity (EC) levels, affecting agricultural productivity.

Electrical conductivity and climate-induced soil salinity

Electrical conductivity (EC) is a crucial parameter reflecting climate-induced soil salinity changes. Prolonged droughts and altered precipitation patterns exacerbate soil salinity issues, impacting electrical conductivity. Pokkali lands, below sea level, experience acid saline conditions due to sea water intrusion, affecting soil fertility and nutrient balance.

Saltwater intrusion and groundwater quality in Munroe Island

Chithra *et al.* (2022) studied salt water intrusion in Munroe Island's groundwater. Increasing sea levels, tidal surges, and reduced river flow elevate salinity, rendering groundwater unsuitable for drinking and irrigation. Climate-induced factors, including sea level rise and changing precipitation patterns, influence salinity values in the Kallada river basin.

Available N, P, K and micronutrients

Effects of short-duration floods on soil Inorganic-N

Unger *et al.* (2009) conducted a study examining the effects of brief floods on soil inorganic nitrogen content, with variations in flood duration (3 weeks vs. 5 weeks) and flow rates (stagnant vs. flowing). The research revealed a decrease in soil NO_3–N levels during 5-week floods, while an increase was observed under 3-week flowing and control conditions. The stagnant treatment exhibited higher losses in soil NO_3–N. Flooded conditions led to NH_4–N production, while NO_3–N was lost, suggesting potential leaching.

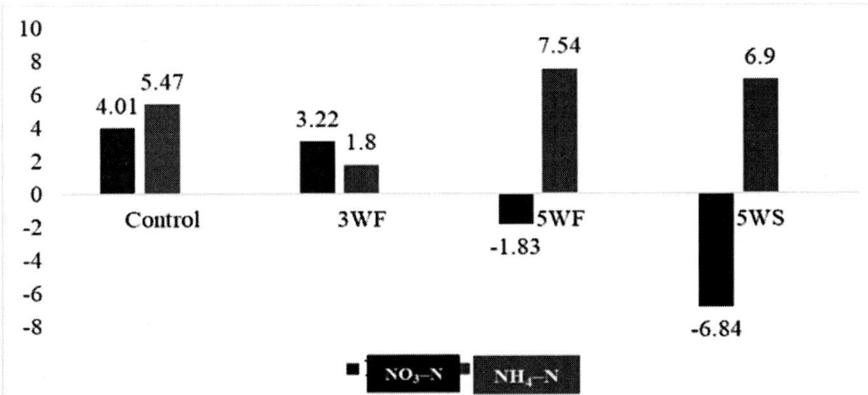

Fig. 2: Average changes in soil inorganic N (mg kg^{-1}) due to flood treatments (Unger *et al.*,2009)

Post-flood soil quality assessment in Idukki, Kerala

Sreekutty (2020) assessed post-flood soil quality in Idukki, Kerala, finding a rise in soil acidity and increases in available phosphorus, potassium, calcium, magnesium, and sulphur. Available boron, however, decreased in post-flood conditions.

Impact of elevated CO_2 on soil nitrogen and phosphorus

Ma *et al.* (2007) conducted a free-air carbon dioxide enrichment (FACE) experiment in a paddy field to study elevated CO_2 effects on soil nitrogen and phosphorus availability. Elevated CO_2 increased rice biomass and phosphorus uptake, while decreasing soil available nitrogen at the tillering stage. Soil available phosphorus decreased initially but increased at later growth stages, suggesting complex interactions under elevated CO_2.

Elevated CO_2 and phosphorus demand

Jin *et al.* (2015) highlighted the likelihood of increased phosphorus demand by plants under elevated CO_2 due to stimulated photosynthesis and altered root exudates. Elevated CO_2 influences root morphology, increases rooting depth, and changes root exudates, impacting phosphorus mobilization in the rhizosphere.

Future soil temperature effects on soil nitrate

Sahoo (2022) studied the potential impact of rising winter soil temperatures on soil nitrate levels using four Representative Concentration Pathways (RCP) scenarios. Increased winter temperatures led to elevated phosphorus, soil pH, and potassium but decreased available nitrogen. A temperature rise of 0.5°C–2.5°C resulted in a significant decrease in available nitrogen, raising concerns about future warming climate responses.

Post-flood soil quality assessment in Kole Wetlands, Kerala

Safnathmol *et al.* (2023) evaluated post-flood soil quality in kole wetlands of Kerala. Post-flood, organic carbon shifted to medium to high levels, and available phosphorus decreased from high to medium. Severe magnesium deficiency and boron deficiency were observed. Available iron reached toxic levels, and Fe, Mn, and Cu content increased due to reducing conditions in paddy fields.

Impact of elevated CO_2 and temperature on soil carbon pools

Samal *et al.* (2020) studied the effects of elevated atmospheric carbon dioxide (CO_2) and temperature on soil carbon pools in a 0–30 cm depth over 5 years. The study revealed significantly lower very-labile and labile carbon fractions in elevated CO_2 and temperature scenarios, indicating a decrease in labile carbon. Conversely, non-labile fractions increased under these scenarios. Active pool (AP) and passive pool (PP) were lower under elevated CO_2 and temperature, highlighting potential consequences for soil fertility and carbon stability in the context of climate change (Stark, 2021).

Table 2: Elevated atmospheric CO_2 and temperature enriched recalcitrant carbon in soil of Sub tropical humid climate (Mg C ha^{-1})

Treatment	Fraction 1 (very labile)	Fraction II (labile)	Fraction III (less labile)	Fraction IV (non- labile)	Active pool (AP)	Passive pool (PP)
Ambient	12.8	7.51	3.6	12.9	20.3	16.5
eCO2	11.5	6.35	1.35	16.6	17.9	17.9
eTemp	10.7	4.99	2.70	14.4	15.7	17.1
eCO_2+eTemp	8.98	6.76	3.49	16.6	15.7	20.1

Source: (Samal *et al.*, 2020)

Cation exchange capacity (CEC)

Nawaz *et al.* (2010) assessed leaching patterns and nutrient losses, specifically potassium (K^+), calcium (Ca^{2+}), and magnesium (Mg^{2+}), in sandy soil with low cation exchange capacity (CEC 2.21 cmol(+) kg^{-1}) and two clayey soils with higher CEC (16.93 and 18.36 cmol(+) kg^{-1}) in the tropics of Thailand. Laboratory experiments mimicking heavy annual precipitation were conducted, simulating three phases of rainy seasons. Sandy soils exhibited higher nutrient leaching, with 2.6%, 10.2%, and 36.6% removal of K^+, Ca^{2+}, and Mg^{2+}, respectively, while clayey soils with elevated CEC showed minimal nutrient concentrations in leachate (Nawaz *et al.,* 2010).

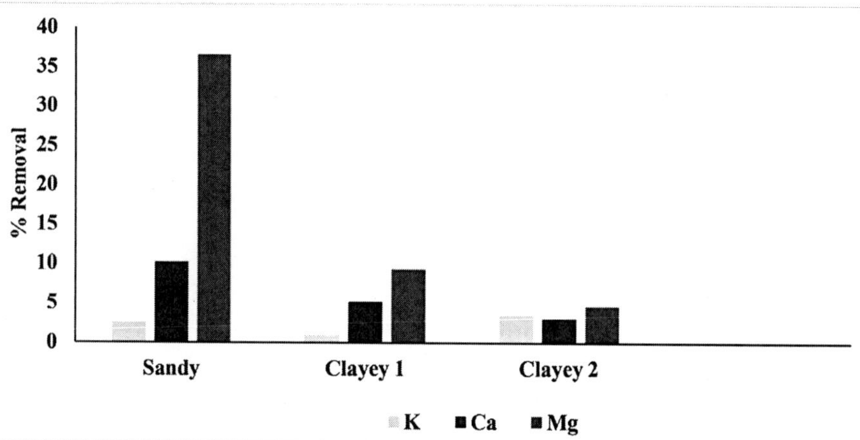

Fig. 3: Percent removal of nutrients from the soils due to leaching in different soil types

Impact of climate change on soil biological properties

Soil organic matter dynamics in Mediterranean Savannahs

San-Emeterio *et al.* (2023) utilized pyrolysis and chemometric analyses to investigate the structural changes in soil organic matter (SOM) in Mediterranean savannahs under five years of rainfall exclusion (drought) and increased temperature (warming). Results revealed a significant decrease in all SOM structural components, especially in open pasture habitats.

Impact of wildfires on soil CO_2 emission

Makhnykina *et al.* (2023) studied the influence of wildfires on soil CO_2 emissions in Central Siberian forests. The research, conducted across different fire chronosequences, demonstrated varying soil CO_2 emissions, with the most recent burn area showing lower emissions compared to sites untouched by fire for 121 years.

Flood-induced changes in microbial diversity in mangroves

Bhat and Bindiya (2019) investigated the impact of floods on microbial diversity in five mangrove ecosystems. Floods altered microbial diversity, with varying phyla dominance in different locations. Mangalavanam showed nine phyla, primarily Proteobacteria, while Puthuvypin exhibited only three phyla, mainly Proteobacteria.

Warming and elevated ozone effects on arbuscular mycorrhizal fungi

Qiu *et al.* (2021) explored the effects of climate warming and elevated ozone on arbuscular mycorrhizal fungi (AMF) in a soybean agroecosystem. Findings indicated reduced root biomass, altered AMF community composition, and decreased AMF colonization under warming and elevated ozone.

Soil acidity impact on earthworms

Chen *et al.* (2020) investigated the effects of soil acidity, induced by acid rain, on earthworms. The study revealed that extreme soil acidity conditions led to earthworm discomfort, direct epidermal damage, and mortality.

Drought impact on soil microbial biomass

Qu *et al.* (2023) assessed the impact of drought on soil microbial biomass carbon (MBC), nitrogen (MBN), phosphorus (MBP), and enzyme activity globally. Drought significantly reduced MBC, MBN, MBP, and enzyme activities on a global scale.

Elevated CO_2 and temperature effects on soil enzymatic activity

Sunoj *et al.* (2015) investigated the influence of elevated CO_2 and temperature on soil enzymatic activity in coconut seedlings. Elevated CO_2 and temperature led to a reduction in dehydrogenase and phosphatase activities in the rhizosphere soil.

Importance of organic agriculture in greenhouse gas emissions

Agricultural practices, including organic agriculture (OA), play a role in the exchange of greenhouse gases (GHGs), such as CO_2, CH_4, and N_2O, between the land surface and the atmosphere. The increase in GHG concentrations contributes to the Earth's temperature rise through radiative forcing (RF). In OA, CO_2 emissions primarily arise from activities like fossil fuel combustion (e.g., farm machinery use) and soil/land-use management practices. Although fossil fuel-derived CO_2 emissions may be lower in OA, there is potential for higher soil CO_2 emissions compared to nonorganic management, as indicated by limited studies. Differences in soil CH_4 emissions may be marginal, with upland OA soils possibly acting as more significant CH_4 sinks, while paddy soils act as stronger CH_4 sources. Livestock CH_4 emissions differences appear to be minimal. However, OA may show smaller soil N_2O emissions than conventional management due to reduced reactive nitrogen inputs. Nonetheless, the use of leguminous cover crops, manure, poultry litter, and higher soil organic carbon stocks in topsoil may contribute to increased N_2O emissions in OA farms. Overall, OA soils may emit 1.05 kg N_2O-N ha^{-1} $year^{-1}$

less than nonorganic management, with a 0.3 kg N_2O-N ha^{-1} $year^{-1}$ difference in temperate regions. While some evidence suggests that OA contributes less to GHG increases, enhancing the database on farm emissions is crucial for making credible comparisons between conventional and OA systems (Lorenz and Lal, 2022).

Organic Agriculture and Climate-Smart Soil Technologies

Soils play a crucial role in the operation of terrestrial ecosystems and the production of food and fiber. One often neglected aspect of soils is their capacity to alleviate greenhouse gas emissions. By following the principles of soil health such as maximize biodiversity, maximize soil cover, minimize disturbance and maximize living roots, a climate smart soil can be created (USDA, 2018). Climate change and GHG mitigation require an 'all of the above' approach, where all reduction measures that are feasible, cost-effective and environmentally sustainable should be pursued. There are different technologies by which climate smart soils are created. This includes bioenergy carbon capture and storage, conservation agriculture, application of organic amendments, integrated nutrient management, integrated farming system and biofertilizer application.

Bioenergy carbon capture and storage

Bioenergy carbon capture and storage, commonly known as BECCS, stands out as a significant topic in discussions about negative emissions technologies. Essentially, this technology combines biopower with carbon capture and storage methodologies previously mentioned. Biomass captures atmospheric CO_2 biologically through photosynthesis during its growth phase, subsequently being used for energy production through combustion. The CO_2 emissions produced during combustion are then captured and stored in appropriate geological reservoirs (Pires, 2019). Agroforestry and Miyawaki represent the primary strategies for implementing bioenergy carbon capture and storage.

Agroforestry

Agroforestry, recognized as a carbon sequestration strategy, was studied by Sureshbhai *et al.* (2017) in Navsari district, Gujarat, India. Four prevalent agroforestry systems, including agrisilvi-horticulture, agrisilviculture, agri-horticulture, and horti-pasture, were evaluated for productivity and carbon sequestration during 2011-12. Eucalyptus in the eucalyptus + spider lily system showed significantly higher woody biomass. Sugarcane under agrisilviculture (teak + sugarcane) system yielded the maximum biomass among intercrops. Total biological yield was higher in agrisilviculture and lowest in agri-horticulture. The agrisilviculture system exhibited the highest carbon sequestration per hectare (tree + intercrop).

Table 3: Carbon sequestration under prevalent agroforestry systems in Gujarat

Agroforestry system	**Woody component (kg tree^{-1})**	**Intercrops (kg m^{-2})**	**Total**	**SOC (%)**
ASHS (Mango + Teak + Brinjal)	5.0	1.98	6.98	0.75
ASS (Eucalyptus + Spider lily)	42.90	4.97	47.87	0.82
ASS (Teak + Sugarcane)	1.5	10.83	12.36	0.44
AHS (Mango + Rice)	1.21	0.73	1.94	0.36
AHS (Mango + Cabbage)	1.37	1.19	2.42	0.57
AHS (Mango + Sapota + Lemon + Coriander)	2.61	0.55	1.93	0.65
HPS (Sapota + Grass)	0.82	0.98	3.5	0.52
CD	1.595	0.473	0.662	-

Source: (Sureshbhai *et al.,* 2017)

Miyawaki

Miyawaki forests, pioneered by Akira Miyawaki, are gaining recognition for climate change mitigation and soil carbon enhancement. These dense and biodiverse forests sequester more carbon per unit area than traditional plantations. With a mix of plant species, Miyawaki forests improve soil health, structure, and fertility, enhancing resilience to climate-induced stressors. Studies in Central Thailand showed Miyawaki forests storing more soil organic carbon compared to nearby ecosystems. In a 100 cm soil depth, Miyawaki forests stored 137.79 t C ha^{-1}, surpassing natural forests, reforested areas, and agricultural soils. Total carbon storage potential in young eco forests using Miyawaki methods was 156.53 t C ha^{-1}, including above and below-ground biomass and soil carbon (Hanpattanakit *et al.,* 2022).

Conservation agriculture

Conservation agriculture (CA) is recognized as an effective approach for controlling soil erosion, enhancing productivity, and improving economic benefits (Ghosh *et al.,* 2015). The four fundamental principles of CA include (i) retaining crop residue mulch; (ii) incorporating cover crops in the rotation cycle;

(iii) using Integrated Nutrient Management (INM) involving a combination of chemical and biofertilizers; and (iv) eliminating soil mechanical disturbances.

In the Indian Sub Himalayas, Ghosh *et al.* (2015) conducted an experiment comparing conservation agriculture with conventional agriculture. The results indicate that adopting suitable Conservation Agricultural (CA) practices, such as using a grass strip of Palmarosa with applied organic amendments (farmyard manure, vermicompost, and poultry manure) along with weed mulching under conservation tillage, enhances system productivity, reduces runoff, soil loss, and conserves soil moisture. CA was found to minimize soil loss by 45 percent and water loss by 50 percent compared to the traditional system.

Mulching, particularly in hydrothermally deficient areas, is a significant production method known for its efficient water use. Investigating the impact of mulching on soil temperature and its potential benefits to crop production, Li *et al.* (2021) conducted three-year field experiments on winter wheat in semiarid rainfed areas. The experiments included three mulching treatments: straw strip mulching (SSM), plastic film mulching (PFM), and non-mulching (CK). The study revealed that, compared with CK and PFM, SSM significantly reduced soil temperature from 15 °C to 13 °C. The findings suggest that SSM can be adopted for sustainable winter wheat production in semiarid rainfed areas.

Organic amendments

Enhancing soil health is most efficiently achieved through the application of various organic manures as soil amendments, particularly in developing countries where organic waste disposal poses a significant challenge (Ros *et al.,* 2003).

According to Geng *et al.* (2022), the utilization of biochar led to a notable increase in soil pH ranging from 8.48% to 79.25%, accompanied by reductions in exchangeable acidity, exchangeable aluminium, and exchangeable H^+ by 56.94% to 94.95%, 34.38% to 95.66%, and 58.72% to 93.27%, respectively. Biochar demonstrated effective capabilities in alleviating oxidative stress in plants, resulting in reduced malondialdehyde and glutathione content in leaves, along with an increase in microbial community diversity. Notably, branch biochar and cow dung biochar exhibited superior soil acidity mitigation effects compared to peanut shell biochar.

Natural fallow

Natural fallow, as a biological process, plays a crucial role in the regeneration of degraded soils and is closely linked to the increase in earthworm population.

When agricultural land is left fallow, it allows for the natural regeneration of vegetation, which, in turn, creates a more favourable habitat for earthworms. Earthworms are essential soil engineers that improve soil structure, aeration, and nutrient cycling. As the fallow period progresses, earthworm populations tend to increase due to the availability of organic matter and improved soil conditions. These earthworms help in breaking down organic materials, which enriches the soil with organic carbon and nutrients. This enhanced biological activity and nutrient availability can contribute to the restoration and improvement of degraded soils, making natural fallow a valuable tool in sustainable land management practices, especially in the context of soil restoration and soil health enhancement.

Tian *et al.* (2000) studied the differences in earthworm populations between natural fallow (Chromolaena), planted fallow and continuous maize-cassava cropping and they found that earthworm population under natural fallow were significantly higher than under planted fallow and continuous maize-cassava because Chromolaena leaves had high N, low polyphenols, intermediate lignin content and maintained consistently higher soil moisture than the planted woody fallow.

Integrated farming system

Integrated Farming Systems (IFS) play a crucial role in sustainable resource management, ensuring food security, and environmental protection. Meera *et al.* (2019) conducted a study on four IFS models—homestead-based, coconut-based, rice-based, and banana-based—to estimate net greenhouse gas (GHG) emissions. With a diverse integration of enterprises such as field crops, horticulture, timber trees, livestock, poultry, fish, and vermicompost, these models demonstrated net negative GHG emissions, except for the rice-based IFS. The incorporation of organic manures, crop residues, and agroforestry components contributed to carbon sequestration, with the homestead-based IFS exhibiting the highest carbon sequestration, while the rice-based IFS showed the lowest. This highlights the effectiveness of IFS in enhancing input use efficiency, fixing carbon into the system, and mitigating GHG emissions.

Biofertilizers

The declining nutritional quality of rice due to rising atmospheric CO_2 levels is currently a significant global concern. Biofertilizers play a crucial role in improving nutrient uptake in plants, especially under climate change conditions such as elevated CO_2, by enhancing the solubilization and availability of essential nutrients in the soil. Bhavya *et al.* (2023) conducted a study to assess the impact of biofertilizers on iron uptake in rice under elevated CO_2.

The experiment included four treatments (KAU, Package of Practices (POP) as a control, POP+Azolla, POP+Plant Growth Promoting Rhizobacteria (PGPR), and POP+Arbuscular Mycorrhizal Fungi (AMF)), each replicated three times under both ambient and elevated CO_2 conditions. The analyzed data revealed that iron uptake and translocation were adversely affected under elevated CO_2, resulting in lower grain quality and reduced iron content. The observed response of iron homeostasis in the experimental plants to the application of biofertilizers, particularly PGPR, under elevated CO_2, strongly suggests the potential use of these biofertilizers in designing iron management strategies to enhance rice quality.

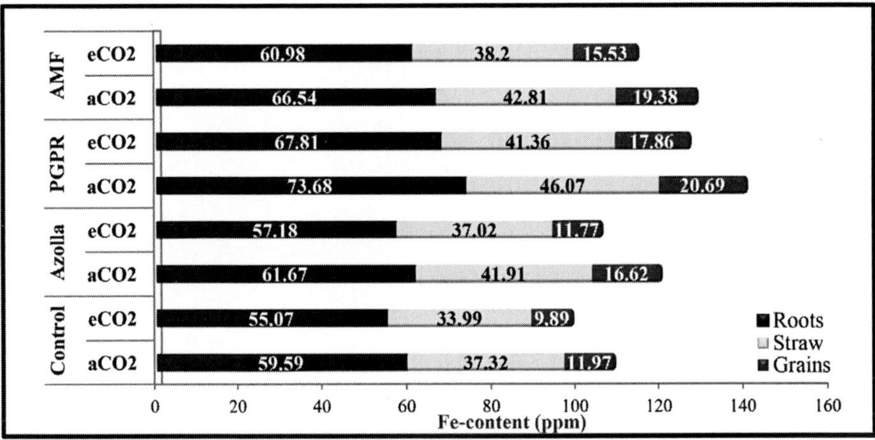

Fig. 4. Iron content in rice plants at harvest under ambient and elevated CO_2 conditions (Bhavya *et al.,* 2023)

Conclusion

Climate change profoundly affects soil health through shifts in temperature, precipitation, and extreme events, impacting erosion, microbial activity, and nutrient cycling. Preserving soil health is crucial for effective climate change mitigation. Healthy soils sequester carbon, enhance water retention, and aid in reducing emissions, contributing to ecosystem resilience. The concept of organic agriculture and climate-smart soils prioritizes adaptive strategies and sustainable practices, aiming to optimize soil use and reduce greenhouse gas emissions. Implementing practices like reduced tillage, precision agriculture, and afforestation can create a cost-effective and sustainable strategy, utilizing soil as a potent ally in combating the challenges of a changing climate.

References

Agbeshie, A.A., Abugre, S., Atta-Darkwa, T., and Awuah, R. 2022. A review of the effects of forest fire on soil properties. J. For. Res. 33(5): 1419-1441.

Alexakis, D., Kokmotos, I., Gamvroula, D., and Varelidis. G. 2021. Wildfre efects on soil quality: application on a suburban area of West Attica (Greece). Geosci. J. 25:243–253

Alirezaei, M., Onat, N.C., Tatari, O., and Abdel-Aty, M. 2017. The climate change-road safety-economy nexus: a system dynamics approach to understanding complex interdependencies. Syst. 5(1): 6.

Archer, D. 2007. Global warming: Understanding the forecast. Blackwell Publishers, Malden, 823p

Bai, M., Impraim, R., Coates, T., Flesch, T., Trouve, R., Van Grinsven, H., Cao, Y., Hill, J., and Chen, D. 2020. Lignite effects on NH_3, N_2O, CO_2 and CH_4 emissions during composting of manure. J. Environ. Manag. 271: 110960.

Bhat, S.G. and Bindiya, E.S.2019. Impact of Flood/Landslides 2K18 on Biodiversity [on-line]. Available: https://doi.org/10.1002/app.49608 [09 Oct. 2023]

Bhavya, M.S.P., Manju, R.V., Viji, M.M., Roy, S., Anith, K.N., and Beena, R. 2023. Impact of bio fertilisers on iron homeostasis and grain quality in the rice variety Uma under elevated CO_2. Front. Plant Sci. 14: 1144905.

Chen, S., Chen, X., and Xu, J. 2016. Impacts of climate change on agriculture: Evidence from China. J. Environ. Econ. Manag.76: 105-124.

Chen, X., Zhang, J., and Wei, H., 2020. Physiological responses of earthworm under acid rain stress. Int. J. Environ. Res. Public Health 7(19): 7246.

Cheng, L., Von Schuckmann, K., Abraham, J.P., Trenberth, K.E., Mann, M.E., Zanna, L., England, M.H., Zika, J.D., Fasullo, J.T., Yu, Y., and Pan, Y. 2022. Past and future ocean warming. Nat. Rev. Earth Environ. 3(11): 776-794.

Chinnasamy, P., Honap, V.U., and Maske, A.B. 2020. Impact of 2018 Kerala floods on soil erosion: Need for post-disaster soil management. J. Ind. Soc. Remote Sensing. 48(10): 1373-1388.

Chithra, S., Joseph, S., and Kannan, N. 2022. A study of saltwater intrusion in the Kallada River, southwest coast of Kerala, India. Water Supply. 22(2): 2194-2211.

Chultheis, E. 2013. Contradicting settled science, Donald Trump says 'nobody really knows' on climate change. The Hindu, 7 Oct. 2013, p.11.

Demirel, Z. 2004. The history and evaluation of saltwater intrusion into a coastal aquifer in Mersin, Turkey. J. Environ. Manag. 70(3): 275-282.

Fauchereau, N., Trzaska, M., Rouault, M., and Richard, Y. 2003. Rainfall variability and changes in Southern Africa during the 20th century in the global warming context. Nat. Hazards 29(2): 139-154.

Geng, N., Kang, X., Yan, X., Yin, N., Wang, H., Pan, H., Yang, Q., Lou, Y., and Zhuge, Y. 2022. Biochar mitigation of soil acidification and carbon sequestration is influenced by materials and temperature. Ecotoxicol. Environ. Saf. 232(1): 113241- 113252.

Ghosh, B.N., Dogra, P., Sharma, N.K., Bhattacharyya, R., and Mishra, P.K. 2015. Conservation agriculture impact for soil conservation in maize–wheat cropping system in the Indian sub-Himalayas. Int. Soil Water Conserv. Res. 3(2):112-118.

Hanpattanakit, P., Kongsaenkaew, P., Pocksorn, A., Thanajaruwittayakorn, W., Detchairit, W., and Limsakul, A. 2022. Estimating carbon stock in biomass and soil of young eco-forest in urban city, Thailand. Chem. Eng. Trans. 97: 427-43

Huang, Y., Zhang, M., Xia, Y., Hu, Y., and Son, S.W. 2016. Is there a stratospheric radiative feedback in global warming simulations?. Clim. Dyn. 46: 177-186.

IMD [India Meteorological Department]. 2021. Weather Reports glossary [on line]. Available: https://www.imdpune.gov.in/ Weather/Reports/glossary org. [05 Oct. 2023].

IPCC [Intergovernmental Panel on Climate Change] 2021. Summary for policymakers. In: Masson-Delmotte, V.P., Zhai, A., Pirani, S.L., Connors, C., Pean, S., Berger, N., Caud, Y., Chen, L., Goldfarb, M.I., Gomis, M., Huang, K., Leitzell, E., Lonnoy, J.B.R., Matthews, T.K., Maycock, T., Waterfeld, O., Yelekçi, R., and Zhou, B. (eds), Climate change 2021: the physical science basis. Proceedings of sixth assessment report of the intergovernmental panel on climate change. Cambridge University Press, Cambridge,pp.52-57.

Jiang, Y. and Khan, H. 2023. The relationship between renewable energy consumption, technological innovations, and carbon dioxide emission: evidence from two-step system GMM. Environ. Sci. Pollut. Res. 30(2): 4187-4202.

Jin, J., Tang, C., and Sale, P. 2015. The impact of elevated carbon dioxide on the phosphorus nutrition of plants: a review. Ann. Bot. 116(6): 987-999.

Kiarsi, M., Amiresmaili, M., Mahmoodi, M.R., Farahmandnia, H., Nakhaee, N., Zareiyan, A., and Aghababaeian, H. 2023. Heat waves and adaptation: A global systematic review. J. Thermal Biol. 103588.

Kleber, G.E., Hodson, A.J., Magerl, L., Mannerfelt, E.S., Bradbury, H.J., Zhu, Y., Trimmer, M., and Turchyn, A.V. 2023. Groundwater springs formed during glacial retreat are a large source of methane in the high Arctic. Nat. Geosci. 16(7): 597-604.

Li, J., Luo, B., Liu, B., Wei, X., Zhong, S., and Wei, C. 2023. Raindrop-impact-induced ejection characteristics of surface particles for soils with a textural gradient. Catena 223: 106930.

Li, Y., Chai, S., Chai, Y., Li, R., Lan, X., Ma, J., Cheng, H., and Chang, L. 2021. Effects of mulching on soil temperature and yield of winter wheat in the semiarid rainfed area. Field Crops Res. 271: 108244.

Lorenz, K. and Lal, R. 2022. Organic Agriculture and Greenhouse Gas Emissions. In Organic Agriculture and Climate Change. Springer International Publishing pp: 129-175.

Ma, H.L., Zhu, J.G., Liu, G., Xie, Z.B., Wang, Y.L., Yang, L.X., and Zeng, Q., 2007. Availability of soil nitrogen and phosphorus in a typical rice–wheat rotation system under elevated atmospheric [CO_2]. Field Crops Res. 100(1): 44-51.

Makhnykina, A., Panov, A., and Prokushkin, A., 2023. The impact of wildfires on soil CO_2 emission in Middle Taiga forests in Central Siberia. Land 12(8): 1544.

Meera, A.V., John, J., Sudha, B., Sajeena, A., Jacob, D., and Bindhu, J.S. 2019. Greenhouse gas emission from integrated farming system models: A comparative study. Green Farming 10(6): 696-701.

Mikhaylov, A., Moiseev, N., Aleshin, K., and Burkhardt, T. 2020. Global climate change and greenhouse effect. Entrep. Sustain. Issues 7(4): 2897.

Miralles, D.G., Teuling, A.J., Van Heerwaarden, C.C., and Vila-Guerau de Arellano, J. 2014. Mega-heatwave temperatures due to combined soil desiccation and atmospheric heat accumulation. Nat. Geosci. 7: 345–349

Mohanty, M.P., Mudgil, S., and Karmakar, S. 2020. Flood management in India: A focussed review on the current status and future challenges. Int. J. Disaster Risk Reduct. 49:101660.

NASA [National Aeronautics and Space Administration]. 2022. NASA Carbon dioxide concentration [on line]. Available: http://climate.nasa.gov/vital-signs/carbon-dioxide/ org. [10 Oct. 2023].

Nawaz, R., Garivait, H., and Anurakpongsatorn, P. 2010. Impacts of precipitation on leaching behavior of plant nutrients in agricultural soils of the tropics. In : Kiasker H.J. and

Mansher, S.T. (eds), Leaching of Nutrients with Precipitation. Proceedings of 2nd International Conference on Chemical, Biological and Environmental Engineering, Singapore. pp. 336-340.

Nayak, S. 2023. Exploring the future rainfall characteristics over India from large ensemble global warming experiments. Clim. 11(5): 94.

NDMA [National Disaster Management Authority]. 2021. Govt. of India- Natural Hazards: Heat Wave [on-line]. Available: https://ndma. gov.in/Natural-Hazards/Heat-Wave. [08 Oct 2023].

Nunez, S., Arets, E., Alkemade, R., Verwer, C., and Leemans, R. 2019. Assessing the impacts of climate change on biodiversity: is below 2° C enough?. Clim. Change. 154: 351-365.

Pires, J.C.M. 2019. Negative emissions technologies: a complementary solution for climate change mitigation. Sci. Total Environ. 672: 502-514.

Prusty, P. and Farooq, S.H. 2020. Seawater intrusion in the coastal aquifers of India-A review. Hydro Res. 3: 61-74.

Qiu, Y., Guo, L., Xu, X., Zhang, L., Zhang, K., Chen, M., Zhao, Y., Burkey, K.O., Shew, H.D., Zobel, R.W., and Zhang, Y. 2021. Warming and elevated ozone induce tradeoffs between fine roots and mycorrhizal fungi and stimulate organic carbon decomposition. Sci. Adv.7(28): 9256.

Qu, Q., Wang, Z., Gan, Q., Liu, R., and Xu, H. 2023. Impact of drought on soil microbial biomass and extracellular enzyme activity. Front. Plant Sci. 14:1221288.

Ros, M., Garcia, C., and Hernandez, T. 2003. Soil microbial activity after restoration of a semiarid soil by organic amendments. Soil Biol. Biochem. 35: 463–469.

Safnathmol, P., Rajalekshmi, K., and Latha, A. 2023. Soil quality assessment and GIS based mapping in post-flood soils of kole wetlands of Kerala. J. Tropic. Agric. 60(2):12-16.

Sahoo, M. 2022. Winter soil temperature and its effect on soil nitrate Status: A support vector regression based approach on the projected impacts. Catena 211: 105958.

Saintilan, N., Kovalenko, K.E., Guntenspergen, G., Rogers, K., Lynch, J.C., Cahoon, D.R., Lovelock, C.E., Friess, D.A., Ashe, E., Krauss, K.W., and Cormier, N. 2022. Constraints on the adjustment of tidal marshes to accelerating sea level rise. Sci. 377(6605): 523-527.

Samal, S.K., Dwivedi, S.K., Rao, K.K., Choubey, A.K., Prakash, V., Kumar, S., Mishra, J.S., Bhatt, B.P., and Moharana, P.C 2020. Five years exposure of elevated atmospheric CO_2 and temperature enriched recalcitrant carbon in soil of subtropical humid climate. Soil Tillage Res. 203: 104707.

San-Emeterio, L.M., Jiménez-Morillo, N.T., Pérez-Ramos, I.M., Domínguez, M.T., and González-Pérez, J.A. 2023. Changes in soil organic matter molecular structure after five-years mimicking climate change scenarios in a Mediterranean savannah. Sci. Total Environ. 857: 159288.

Scambos, T. and Moon, T. 2022. How is land ice changing in the Arctic, and what is the influence on sea level?. Arct. Antarct. Alp. Res. 54(1): 200-201.

Sreekutty, M.R. 2020. Assessment of soil quality in the post flood scenario of AEU 16 in Idukki district of Kerala and generation of GIS maps. M.Sc. (Ag) Thesis, Kerala Agricultural University, Thirssur, 82p.

Stark, A.M. 2021. Lab researchers find elevated CO2 emissions increase plant carbon uptake but decrease soil carbon storage [on-line]. Available: https://www.llnl.gov/article/47406/lab-researchers-find-elevated-co2-emissions-increase-plant-carbon-uptake-decrease-soil-carbon [06 Oct. 2023]

Sunoj, V.S., Kumar, S.N., Muralikrishna, K.S., and Padmanabhan, S. 2015. Enzyme activities and nutrient status in Coconut (Cocos nucifera L.) seedling rhizosphere soil after exposure to elevated CO_2 and temperature. J. Indian Soc. Soil Sci. 63(2): 191-199.

Sureshbhai, P.J., Thakur, N.S., Jha, S.K., and Kumar, V. 2017. Productivity and carbon sequestration under prevalent agroforestry systems in Navsari District, Gujarat, India. Int. J. Curr. Microbiol. Appl. Sci. 6(9): 3405-3422.

Taylor, C., Robinson, T.R., Dunning, S., Rachel Carr, J., and Westoby, M. 2023. Glacial lake outburst floods threaten millions globally. Nat. Commun. 14(1): 487.

Unger, I.M., Motavalli, P.P., and Muzika, R.M. 2009. Changes in soil chemical properties with flooding: A field laboratory approach. Agric. Ecosyst. Environ.131(1-2): 105-110.

USDA [United States Department of Agriculture] 2018. Principles, Requirements, and Guidelines for Water and Land Related Resources Implementation Studies [on-line] Available: PR&G+5-10-2018+final%20(5).org. [09 Oct. 2023]

USEPA [United States Environmental Protection Agency] 2023. Environmental Justice [on-line].Available: https://www.epa.gov/environmentaljustice. [06 Oct. 2023]

Verma, S. and Jayakumar, S. 2018. Effect of recurrent fires on soil nutrient dynamics in a tropical dry deciduous forest of Western Ghats. India J Sustain. 37(7):678–690

Vijayavenkataraman, S., Iniyan, S., and Goic, R. 2012. A review of climate change, mitigation and adaptation. Renew. Sust. Energ. Rev. 16(1): 878-897.

Wadhawan, S.K., Singh, B., and Ramesh, M.V. 2020. Causative factors of landslides 2019: case study in Malappuram and Wayanad districts of Kerala, India. Landslides. 17: 2689-2697.

WMO [World Meteorological Organization] 2020. The State of the Global Climate 2020. Available: https://public.wmo.int/en/our-mandate/climate/wmo-statementstate-ofglobal - climate [10 Oct. 2023].

Yashmin, N., Jamuda, M., Panda, A.K., Samal, K., and Nayak, J.K. 2022. Emission of greenhouse gases (GHGs) during composting and vermicomposting: Measurement, mitigation, and perspectives. Energy Nexus 7: 100092.

Zare, M., Azam, S., and Sauchyn, D. 2022. Impact of climate change on soil water content in southern Saskatchewan, Canada. Water 14(12): 1920.

Tian, G., Olimah, J. A., Adeoye, G. O., and Kang, B. T. 2000. Regeneration of earthworm populations in a degraded soil by natural and planted fallows under humid conditions. Soil Sci. Soc. Am. J. 64(1): 222-228.

4

Mitigation and Adaptation Strategies for Climate Change and Physiology of Plants

Sneha Hajare, Arindam Deb, Asha Sastya and Mayank Pratab Singh Bangari

Department of Plant Physiology, College of Agriculture, Vellayani

Abstract

Climate change is the term used to describe fluctuations in the average climate that might arise as a result of either natural or human-induced processes that have the potential to affect every living organism. The global carbon dioxide concentration (CO_2) has begun to rice since the time of industrial revolution due to human induced climate change activity since 1950. The Intergovernmental Panel on Climate Change reports that the global surface temperature increased by 1.09°C between 2011 and 2020 compared to 1850–1900 m. In 2023, the temperature was recorded at 1.5°C above preindustrial levels. This increase is expected to continue until mid-century, intensifying variations in the global water cycle, global monsoon precipitation, and the intensity of wet and dry events in all ecosystems (IPCC 2023). Climate change has multidimensional and complicated interactions between CO_2 concentration, temperature, water availability, and nutrient availability that making it detrimental to understand how and which crops are affected by each or both temperature variations and in which developmental stage crops are most affected that ultimately affect crop output. Temperature and carbon dioxide (CO_2) are the two main environmental variables that have been demonstrated to have increased significantly over the past century and are directly linked to changing plant growth and development. Therefore, the immediate need for the development of mitigation strategies over climate change to improve crop performance in order to adapt to adverse and frequently varying climatic conditions is of paramount importance. The impacts of elevated CO_2 and temperature on the key physiological processes such as photosynthesis, respiration, transpiration and nitrogen metabolism are discussed in this section.

Keywords: *Climate change, Photosynthesis, Elevated CO_2, Crop productivity and Nitrogen metabolism*

Introduction

Climate change has significant adverse implications on the physiological processes and metabolic machinery of plants, potentially affecting growth, development, and overall crop productivity. The rapid increase in global population exacerbates the threat to global food security, which is further impacted by changing climate conditions. The Intergovernmental Panel on Climate Change (IPCC) estimates that between 1850 and 1900 and 2011 and 2020, the global mean land surface air temperature increased by 1.6 °C. Predictions by the United Nations indicate that the global human population could reach 8.5 billion by 2023 and 9.7 billion by 2050, up from the current 8 billion. The National Centers for Environmental Information (NCEI) reported that 2022 was 0.86°C warmer than the average of the 20th century, making it the 6th warmest year on record. These data suggest an increased probability of unpredictable climate changes and harsh growing conditions, manifesting as various abiotic stresses on major cultivable lands. Climate change impacts CO2 concentration, temperature, precipitation, pollutants, and nutrient availability and uptake, leading to drought, salinity, irregular rainfall, flooding, high temperatures, and nutrient deficiencies in staple food crops, thereby threatening global food security. This review focuses on the effect of elevated CO2 and temperature on fundamental plant metabolic processes such as photosynthesis, respiration, transpiration, and nitrogen metabolism.

The effects of climate change on Physiological processes of crops

Climate change has significant adverse implications on the physiological processes and metabolic machinery of plant that can have potential negative affect on growth, development, and overall crop productivity. Rapid increase in global population has imposed an additional threat to the global food security which is further, significantly affected by the changing climate. The Intergovernmental Panel on Climate Change (IPCC) estimates that between 1850 and 1900 and 2011 and 2020, the global mean land surface air temperature increased by 1.6 °C. According to the United Nations prediction on the global human population, it could reach 8.5 billion by 2023 and 9.7 billion by 2050 from the present 8 billion population. National Centres for Environmental Information (NCEI) reported that 2022 was 0.86°C (1.55°F) warmer than the average of the 20th century by a temperature of 13.9°C (57.0°F), making it the 6th warmest year ever since the global records began since the year 1880.

These data suggests that there is an increased probability of occurrence of unpredictable climate changes along with increased intensity and frequency of harsh growing conditions in the form of various abiotic stresses on the major cultivable lands of the world in the future days are the inevitable phenomenon. Climate change causes variation in the CO_2 concentration, temperature, precipitation, pollutants, and nutrient availability and uptake which results in the occurrence of drought, salinity, irregular rainfall, flooding, high temperature, elevated CO_2 and nutrient deficiency in the major staple food crops threatening the global food security. Hence it is of prime importance to address the effect of changing climatic conditions on the physiological processes of the plants. This review mainly focuses on the effect of elevated CO_2 and temperature on the basic metabolic machinery of the plants such as photosynthesis, respiration, transpiration, and nitrogen metabolism.

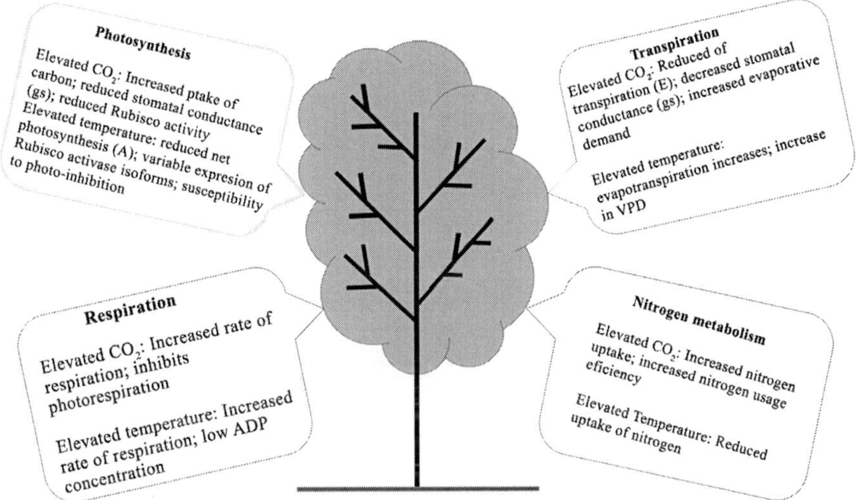

Fig. 1: Effect of climate change on physiological processes.

The effect of climate change on photosynthesis and allied traits:

The effect of elevated CO_2 on photosynthesis

The atmospheric concentration of CO_2 has escalated from 280 μmol^{-1} to 410 μmol^{-1} due to industrialization (Ciais *et al.*, 2013). According to the data published by Global Monitoring Laboratory suggests that the atmospheric concentration of CO_2 has reached 421.86 ppm in the January of 2024. It is predicted that the CO_2 concentration will elevate two-fold, i.e., up to 800 μmol^{-1} at the end of present century. The rate of CO_2 uptake and release by terrestrial plants through photosynthesis and respiration, respectively, will have a major impact on how the atmospheric CO_2 concentration changes over time. These increased CO_2 concentration in the atmosphere will affect several physiological processes, growth and yield parameters of the plants. It includes increase in the rate of photosynthesis (Drake *et al.,* 1997; Kaiser *et al.,* 2017; Rakhmankulova *et al.,* 2023), leaf area and tiller development that will ultimately effect the plant biomass production and yield of major crops (Costa *et al.,* 2003; Sekhar *et al.,* 2023). C3 being a widely adapted mechanism of photosynthesis, and since CO_2 is one of the limiting factors for the photosynthesis process, elevated CO_2 conditions results in increases the carbon uptake (Fig.1) (Furbank & Taylor, 1995; Sekhar *et al.,* 2023), by improving the carboxylation activity and preventing the oxygenation reaction of rubisco, glycolate production, and CO_2 released in the photo respiratory phase (Long & Ort 2010). Long-term exposure to increased CO_2 conditions can preserve the initial stimulation of net photosynthesis (Idso & Kimball 1992; Gunderson, Norby & Wullschleger 1993; Teskey 1995), although it is frequently completely reversed or eliminated (Stitt 1991; Oechel *et al.* 1993; Sage 1994; Nie *et al.* 1995a; Jacob, Greitner & Drake 1995; Tissue, Thomas & Strain 1997; Drake *et al.* 1997). This process is termed as 'acclimation' of photosynthesis to elevated CO_2.

Acclimation is usually characterized by alterations in the changes in the gas exchange parameters that are signifying the reduction in carboxylation capacity (Stitt 1991; Sage 1994) and decrease of Rubisco activity or Rubisco protein content in many species (Drake *et al.,* 1997; Park *et al.,* 2023), such as *Arabidopsis thaliana* (Cheng, Moore & Seeman 1998), tobacco (Sicher & Kremer 1994; Sicher, Kremer & Rodermel 1994), tomato (Van Oosten, Wilkins & Beresford 1994), wheat (Nie *et al.,* 1995a; Rogers *et al.,* 1996; Sicher & Bunce 1997), *Scirpus olneyi* (Jacob *et al.,* 1995), barley (Sicher & Bunce 1997), rice (Vu *et al.* 1997), *Panicum laxum* (von Ghannoum *et al.,* 1997), loblolly pine (Tissue, Thomas & Strain 1993; Tissue *et al.,* (1997), cherry (Atkinson *et al.*, 1997), oak (George, Gerant & Dizengremel 1996) and cork oak (Faria *et al.,* 1996). Long term exposure of plants to elevated CO_2 has also been demonstrated to lower the level of transcript for various proteins involved in photosynthesis. The process of acclimation of photosynthesis to elevated CO_2 is due to the suppression of Rubisco activity (Fig.1) and other key proteins that are essential for photosynthesis (Stitt 1991; Jang & Sheen 1994; Van Oosten & Besford 1996). Although the levels of several transcripts encoded by plastids, such as RbcL, PsaA, PsaB, and PsbA (encoding key proteins in photosystems I and II and the larger subunit of Rubisco), do not decrease under elevated CO_2, the transcripts for the small subunit of Rubisco encoded by nucleus (RbcS), chlorophyll binding protein (Cab), and Rubisco activase (Rca) enzyme decrease in elevated CO_2 conditions in tomato (Van Oosten *et al.,* 1994; Van Oosten & Besford 1995). RbcS, RbcL, and three other Calvin cycle enzyme transcripts decreased in wheat in response to increased CO_2 (Nie *et al.* 1995b). Long term exposure to increased CO_2 causes Arabidopsis to have lower levels of RbcS and RbcL transcripts (Cheng *et al.,* 1998). Elevated level of CO_2 in wheat, bean, parsley, bugle, cotton, sunflower, and radish has lowered the amount of RbcS transcript (Moore *et al.,* 1998).

Plants cultivated under higher CO_2 conditions have demonstrated better development and photosynthesis, lower rate of transpiration, improved water use efficiency, decreased inorganic nutrition concentration, increased plant hormone, and compact stomatal density than the plants grown under ambient CO_2 conditions. However, this response is not linear (Farquhar *et al.,* 1980). The rate of photosynthesis will be stimulated when the Ci is low compared to the conditions where the Ci is high, (Farquhar *et al.,* 1980; Sharkey *et al.,* 2007) since the stomatal conductance under such conditions will be low and CO_2 diffusion into the leaf is restricted (Ainsworth & Rogers, 2007). The regeneration of RuBP and the rate of use triose phosphate to synthesize starch and sucrose will limit the rate of photosynthesis (Sharkey *et al.,* 1986). This indicates that the increase in the concentration of CO_2 that occurred since the Industrial Revolution has resulted in the increase in the rate of photosynthesis (Polley *et al.,* 1993; Gerhart & Ward, 2010). However, further increase in the CO_2 in the future days might impose lower effect of the rate of carbon accumulation which is consistent with the saturated response of elevated CO_2 with crop yield studies (Long *et al.,* 2006). Stomatal conductance have already adapted to the rise CO_2 with reduction in the maximum gs over the last century via decreases instomatal density or stomatal pore size (Lammertsma *et al.,* 2011).Further the over-accumulation of the sugars under elevated CO_2 conditions results in negative feedback response on phototsynthesis due to Rubisco down-regulation (Moore *et al.,* 1999). Elevated CO_2 has reduced stomatal conductance (gs), (Medlyn *et al.,* 2001) by 20% across the FACE experiments (Ainsworth & Long, 2004) although in some studies, this reduction of gs disappeared over time (Uddling *et al.,* 2009) or was not seen (Pathare *et al.,* 2017). Since the increased resistance to CO_2 diffusion into the leaves lowers the C_i, it results in reduced stimulation of Rubisco carboxylation rates.

Most of the elevated CO_2 studies have been conducted under controlled growth chambers and glasshouse conditions where the crops are maintained under controlled ambient nutrition, water condition in the pots (Mcconnaughay *et ul.,* 1993; Qiu *et al.,* 2023). There has been a significant difference between the pot and field experiemnts (Poorter *et al.,* 2016; Hassan *et al.,* 2023). Therefore, the realistic setting for the elevated CO_2 experiments includes open-top chambers (Long *et al.,* 2023) and Free Air CO_2 Enrichment (FACE) experiments (Yang *et al.,* 2023). It involves the exposure of the plants grown under natural field condition to a blowing elevated CO_2 condition (Rogers *et al.,* 2004; McCarthy *et al.,* 2010; Smith *et al.,* 2014; Obermeier *et al.,* 2017). Maximum rate of assimilation was recorded to be 31% among the FACE experiments (Ainsworth & Long, 2004).

The effect of elevated temperature on photosynthesis

Heat stress is characterized as the rise in temperature above a threshold point for a sufficient period of time to cause irreparable damage to plant growth and development (Wahid et al., 2006). Heat stress will become more intense by 2.7°C and occur more frequently by 13.9 times in a future with a 2.0°C warming level in a climate unaffected by human activity than it did between 1850 and 1900 (IPCC, 2021). The response of photosynthesis to high seasonal temperatures varies depending on the crop. Plants can achieve metabolic homeostasis through the process of thermal acclimation that includes adjusting metabolic rates to compensate for variations in the temperature. As leaf temperature rises, net photosynthesis increases, peaks at an optimum temperature, and subsequently declines (Posch *et al.,* 2019). Long-term effects of global warming could result in photosynthetic acclimation through improved electron transport capacity of photosystem II (PSII) (Yamasaki *et al.*, 2002), variable expression of Rubisco activase isoforms, expression of heat shock protein (HSP), variation in susceptibility to photo-inhibition (Hurry and Huner, 1991, 1992; Oquist *et al.*, 1993) and respiration reduction; however, this could continue to contribute to lower photosynthetic rates of the crops compared to those at ambient temperature conditions (Way & Yamori, 2014; Yamori, Hikosaka, & Way, 2014). Heat waves and acute heat stress, which are characterized by abrupt increases in temperature, reduce the possibility of photosynthetic acclimation of the crops (Smith and Dukes 2017). For example, field-grown maize and soybeans grown during the heat wave for three-day showed significant decrease in leaf photosynthesis. However, within 24 hours of the recovery period, photosynthesis revived back to control levels (Siebers *et al.*, 2015, 2017). While simulated heat waves usually resulted in lower crop productivity, the effects were greater when heating coincides with the reproductive stages of the crop (Siebers *et al.*, 2015, 2017).

The catalytically sluggish enzyme rubisco is an essential component in the carbon fixation process. It is involved in carboxylation process with considerable affinity towards CO_2 and oxygenation processes which follows photorespiratory pathway to recycle the inhibitory by-products. The concept of Rubisco's trade-off between specificity and catalytic activity hinders attempts to design an enzyme with higher efficiency. Favouring the carboxylation process of the RuBP enzyme is crucial for improving the carbon fixation under current warmer climate conditions. First, under warmer temperatures, Rubisco's catalytic activity, specificity, and efficiency can be increased toward increased CO_2 (Cavanagh *et al.,* 2023). Secondly, by enhancing the expression of the Rubisco activase isoforms through the utilization of wild varients, the rubisco activase enzyme's thermotolerance can be improved. Thirdly, increase in the concentration of the rubisco enzyme can be achieved by increasing the amount of caroxysomes and convertion of CO_2 sensitive C3 to CO_2 concentration independent C4 plants. Finally, increased expression of the core enzymes involved in the RuBP regeneration could result in improved carboxylation of rubisco enzyme (Slattery and Ort 2019).

The effect of climate change on respiration

The effect of elevated CO_2 on respiration

The carbon flux created from the vegetation is higher than the CO_2 created by the human activities (Le Quere *et al.,* 2016), it could results in either mitigation or acceleration of climate change (Smith & Dukes, 2013). Carbon flux in the surrounding determines the balance between the photosynthesis and respiration. These higher CO_2 concentrations, along with increasing concentrations of other greenhouse gases, have led to a 0.8°C rise in mean annual global temperatures as of 2017 (Hansen *et al.,* 2010). Higher CO_2 level has no significant effect on the mitochondrial respiration (Amthor, 2000). However, Wang *et al.,* (2001) has showed that the plants grown under elevated CO_2 showed higher rate of respiration, contrastingly, Curtis, 1996 have shown that the rate of respiration declines. High CO_2 inhibits photorespiration (Fig.1), which is typically considered to be a wasteful consequence of Rubisco.

The effect of elevated temperature on respiration

The two primary temperature-sensitive processes that determine the carbon balance are photosynthesis and respiration. The rate of respiration rises with temperature up to a peak at Tmax, beyond which it declines with further temperature increases. At lower measurement temperatures, the hot-acclimated plant exhibits lesser respiration compared to the cold-acclimated plant. Respiration starts to decrease at lower temperatures in the cold-acclimated plant than in the hot-acclimated plant because Tmax correspondingly occurs at a higher temperature in the hot-acclimated plant (Posch *et al.,* 2019). Around the time that the condition of the mitochondrial membrane affected the activity of alternative oxidase in germinating wheat, alternative respiration reached its peak. Additionally, at this point, respiration dropped in part because oxygen became less soluble at higher temperatures (McCaig *et al.,* 1977). Warmer temperatures cause mitochondrial swelling and downregulation of respiration at the cellular level. This is attributed to various factors such as low ADP concentration or increase in the ATP:ADP ratio, a reduction in ATP transfer to the cytosol mediated by abscisic acid (Berkowitz *et al.,* 2016), disruption of the concentration gradient of intermediates in the tricarboxylic acid cycle (TCA cycle) (Atkin *et al.,* 2003), lipid peroxidation in mitochondrial membranes (Welchen *et al.,* 2016), and cytochrome c release (Zancani *et al.,* 2015; Petereit *et al.,* 2017), which in turn causes programmed cell death (Sharma *et al.,* 2022; Aken and Breusegem *et al.,* 2015; Pedrotti *et al.,* 2018).

The effect of climate change on transpiration

The effect of elevated CO_2 on transpiration

Improved physiological understanding of the effect of rising atmospheric CO_2 concentrations on plant transpiration is required. There are various mechanisms involved with several undefined reasons related to the complex physiological and radiative processes involved with plant transpiration under elevated atmospheric CO_2 concentration. Raising CO_2 will have significant effects on future climate, global vegetation, and water availability, all of which are necessary to maintain the productivity of terrestrial ecosystems. Plant physiological processes, atmospheric evaporative demand, water availability, emerging plant disturbances linked to rising temperatures, and changes in plant physiology and coverage are anticipated to drive future changes in global plant transpiration in response to increased CO_2.

A rise in the environment CO_2 concentration may result in decreased stomatal conductance (gs) due to a decrease in stomatal aperture and frequency (Morison 1987; Field, Jackson & Mooney 1995; Drake *et al.,* 1997). Reductions in stomatal conductance (gs) results in reductions in the rate of transpiration (E). Reduced transpiration rate (E) raises the temperature of the leaves by causing a partial imbalance in the saturation of the air surrounding the vegetation. This decreases the evaporative cooling capacity of the leaves resulting in increased evaporative demand of the crops. Reduced transpiration over a period of time can lead to improved soil moisture availability even in the absence of changes in leaf area (Field *et al.,* 1995). Improved soil moisture availability for transpiration would allow the maintenance of gs and hence increases rate of assimilation (AN) (Knapp Hamerlynck & Owensby 1993; Jackson *et al.,* 1994; Owensby *et al.,* 1997). The imposed reductions in transpiration may be partially countered by increases in the amount of carbon available for new leaves to grow. (Heath & Kersteins 1997).

The effect of elevated temperature on transpiration

The process of transferring water from the soil surface to the atmosphere, known as evapotranspiration, necessitates transforming the phase of the water from liquid to vapour (Wang *et al.,* 2012). Numerous factors, including as soil and crop characteristics, meteorological conditions, management practices, and environmental variables including salinity, heat, and water, influence the amount of evapotranspiration (Qiu *et al.,* 2017; Minhas *et al.,* 2020; Yang *et al.,* 2020). Crop evapotranspiration is primarily affected by heat stress. The impact of heat stress on evapotranspiration is getting increased attention as a result of the rising frequency and severity of heat stress globally (IPCC, 2021). Since, heat stress is a direct result of the ongoing global warming conditions, it is important to devote more attention to the ways by which it affects evapotranspiration.

The daily evapotranspiration increased due to heat stress. Under high temperature conditions, increased crop evapotranspiration in wheat was observed, which resulted in decreased soil moisture in later seasons (Asseng *et al.,* 2011). Only during the short heat stress phase of the vegetative growth stage could the daily transpiration recover. When heat stress persisted for a longer period of time, the drop in the increment of rate of daily evapotranspiration was observed. The evaporation rate increased by 100%–115% during the heat stress period due to an increase in Vapor pressure deficit (VPD) (Fig.1) caused by heat stress, which also raised the potential evaporation rate in gerbera (Yang *et al.,* 2023). However, following heat stress, daily evapotranspiration decreased throughout the recovery phase. Under high temperature stress, a considerable increase in plant transpiration rates results in a trade-off between the demand for cooling and water conservation (Sadok *et al.,* 2021).

The effect of climate change on nitrogen metabolism

The effect of elevated CO_2 on nitrogen metabolism

Nitrogen content generally decreases in plants grown under elevated CO_2 conditions (Gojon *et al.,* 2023; Liu *et al.,* 2023). Elevated CO_2 leads to increased nitrogen uptake, provided if there is enough nitrogen available for the plant. Nitrogen-limited plants exhibits a greater degree of photosynthetic acclimation to elevated CO_2 than well-fertilized plants. The shift to a reduced carboxylation capacity under elevated CO_2 is more pronounced in nutrient-limited plants compared to well-fertilized plants, according to the gas exchange characteristics (Arp 1991; Sage 1994). Additionally, elevated CO_2 causes a greater decrease in Rubisco content in nitrogen limited plants than in well-fertilized plants (Riviere-Rolland, Contard & Betsche 1996; Rogers *et al.,* 1996; Ziska *et al.,* 1996). Increased CO_2 caused the RbcS transcript level and Rubisco activity in pea plants cultivated in limited nitrogen to decrease, but not in plants grown in limiting phosphate or with a high nitrogen supply. Due to their slower growth rates, the latter treatment indicates the decrease in Rubisco content in the nitrogen limited plants was additionally having a non-specific impact (Riviere-Rolland *et al.,* 1996).

Increased non-structural carbohydrate content is usually the result of elevated CO_2 (Stitt 1991; Zheng *et al.,* 2019). Additionally, the relative amount of starch and sugar accumulation in elevated CO_2 are influenced by the nitrogen supply. An elevated concentration of carbon dioxide resulted in a significant rise in sugar levels, while there was no substantial increase in starch when *Nicotiana plumbaginifolia* cultivated hydroponically with high nitrate. Conversely, a notable increase in starch and a comparatively smaller increase in sugars was observed when the plants were grown in pots with lower nitrogen supply, (Ferrario-Mery *et al.,* 1997). Similar results were observed in sugar beet (Kutik *et al.,* 1995) and *Pinus palustris* (Pritchard *et al.,* 1997).

The growth rate per unit of nitrogen uptake in the plants, or nitrogen usage efficiency (Fig.1), has increased in elevated CO_2 conditions (McKee and Woodward 1994; Jacob *et al.,* 1995: Kant *et al.,* 2012). The increase in the nitrogen use efficiency is due to the reduced nitrate content under elevated CO_2 conditions (Hocking and Meyer 1985: Cruz *et al.,* 2003). In pea and wheat, the amount of other leaf proteins increase in tandem with the decrease in Rubisco under elevated CO_2 (Riviere-Rolland *et al.,* 1996; Rogers *et al.,* 1996). Alternatively, the nitrogen released by Rubisco and other proteins involved in photosynthetic system could be directed toward the growth of non-photosynthetic organs.

The effect of elevated temperature on nitrogen metabolism

In plants, nitrogen metabolism is intimately linked to both photosynthesis and respiration. The processes of nitrogen metabolism include assimilation, transport, metabolism of amino acids, and absorption of nitrogen (Kusano *et al.,* 2011). Plants have the ability to take up inorganic nitrogen (NO^{-3} and NH^{+4}) from the soil. Once NO^{-3} enters the cell, nitrate reductase (NR) reduces it to NO^{-2}, and nitrite reductase (NiR) reduces NO^{-2} to NH^{+4}. NH^{+4} is subsequently incorporated via the glutamine-glutamate cycle pathway into amino acids and other organic materials (Liang *et al.,* 2021) which uses energy for metabolism.

However, a number of environmental stressors can cause plants to have inadequate energy metabolism, which further impairs the plants' ability to absorb inorganic nitrogen. The uptake of nitrogen is severely threatened by heat stress conditions, which is one of a common environmental issue that can have an impact on plant development and productivity (Gaudreau *et al.*, 1995). Heat-tolerant plants' leaves showed increased activity of NR, glutamine synthase (GS), glutamate synthase (GOGAT), and glutamate dehydrogenase (GDH) in response to heat stress, but the amount of amino acids in the leaves overall remained unchanged (Yuan *et al.*, 2017).

Conclusion

Global mean temperature and atmospheric carbon dioxide (CO_2) are expected to rise sharply by the end of the twenty-first century. Plant physiology and growth of the plants are influenced by both elevated CO_2 and higher temperatures. Future plants are likely to be subjected to higher concentrations of CO_2 as well as more intense heat stress, which can have significant detrimental effects on ecosystem productivity and biodiversity. The effects of increasing CO_2 on plant physiology and growth vary according on the temperature regime, which has major implications for biomass accumulation, rubisco activity, and ecosystem functioning when CO_2 levels rise and climate extremes occur in the future. Because CO_2 is less soluble than O2 and Rubisco is less specific for CO_2 at higher temperatures, photorespiration is increased by increasing temperatures. Therefore, it has been anticipated that at higher temperatures, photosynthetic response to elevated CO_2 in plants with C3 metabolism will be greater. On the other hand, due to their CO_2 concentration mechanism, it has been widely accepted that C4 species will exhibit minimal CO_2 stimulation regardless of temperature. However, under non-HS conditions, recent research has demonstrated a significant stimulation of biomass and net photosynthesis in C4 species. Though much research has been done, little is known about the independent and combined effects of temperature and elevated CO_2 on the physiology and growth of many different plant species.

References

Ainsworth EA, Long SP. 2004.What have we learned from 15 years of free-air CO2enrichment (FACE)? A meta-analytic review of the responses of photosynthesis, canopy properties and plant production to rising CO2.New Phytologist165: 351–372.

Ainsworth EA, Rogers A. 2007.The response of photosynthesis and stomatal conductance to rising [CO2]: mechanisms and environmental interactions. Plant, Cell & Environment30: 258–270.

Amthor JS. 2000. Direct effect of elevated CO2 on nocturnal in situ leaf respiration in nine temperate deciduous tree species is small. Tree Physiology20: 139–144.

Arp W.-J. (1991) Effects of source-sink relations on photosynthetic acclimation to elevated carbon dioxide. Plant, Cell and Environment 14, 869–876.

Asseng, S., Foster, I.A.N. and Turner, N.C., 2011. The impact of temperature variability on wheat yields. Global Change Biology, 17(2), pp.997-1012.

Atkin, O.K. and Tjoelker, M.G., 2003. Thermal acclimation and the dynamic response of plant respiration to temperature. Trends in plant science, 8(7), pp.343-351.

Atkinson C.J., Taylor J.M., Wilkins D. & Besford R.T. (1997) Effects of elevated carbon dioxide on chloroplast components, gas exchange and growth of oak and cherry. Tree Physiology 17, 319–325.

Berkowitz, O., De Clercq, I., Van Breusegem, F. and Whelan, J., 2016. Interaction between hormonal and mitochondrial signalling during growth, development and in plant defence responses. Plant, cell & environment, 39(5), pp.1127-1139.

Cavanagh, A.P., Slattery, R. and Kubien, D.S., 2023. Temperature-induced changes in Arabidopsis Rubisco activity and isoform expression. Journal of experimental botany, 74(2), pp.651-663.

Cheng S.-H., Moore D. & Seeman J.R. (1998) Effects of short- and long-term elevated carbon dioxide on the expression of ribilose1,5-bisphosphate carboxylase/oxygenase genes and carbohydrate accumulation in leaves of Arabidopsis thaliana (L.) Heynh. Plant Physiology 116, 715–723.

Ciais P, Sabine C, Bala G, Bopp L, Brovkin V, Canadell J, Chhabra A, DeFries R,Galloaway J, Heimann Metal.2013.Carbon and other biogeochemical cycles. In: Tignor M, Allen SK, Boschung J, Nauels A, Xia Y, Bex V, Midgley PM, eds. ClimateChange2013:thePhysicalScienceBasis.ContributionofWorkingGroupItothe Fifth Assessment Report of the Intergovernamental Panel on Climate Change.Cambridge University Press, Cambridge, UK and New York, NY, USA, 465–570.Hansen J, Ruedy R, Sato M, Lo K. 2010. Global surface temperature change. Reviews of Geophysics 48: RG4004.

Costa, W.A.J.M., W.M.W. Weerakoon, H.M.L.K. Herath, R.M.I. Abeywardena. 2003. Response of growth and yield of rice (Oryza sativa L.) to elevated atmospheric carbon dioxide in the subhumid zone of Sri Lanka. J. Agron. Crop Sci., 95 (189): 83-95.

Cruz, C., Lips, H. and Martins-Loução, M.A., 2003. Nitrogen use efficiency by a slow-growing species as affected by CO2 levels, root temperature, N source and availability. Journal of plant physiology, 160(12), pp.1421-1428.

Curtis PS. 1996.A meta-analysis of leaf gas exchange and nitrogen in trees grownunder elevated carbon dioxide.Plant, Cell & Environment19: 127–137.

Drake B.G., Gonzàlez-Meler M.A. & Long S.P. (1997) More efficient plants: a consequence of elevated carbon dioxide? Annual Review of Plant Physiology and Plant Molecular Biology 48, 607–640.

Faria T., Wilkins D., Besford R.T., Vaz M., Pereira J.S. & Chaves M.M. (1996) Growth at elevated carbon dioxide leads to down regulation of photosynthesis and altered response to high temperature in Quercus suber seedlings. Journal of Experimental Botany 47, 1755–1761.

Ferrario-Mery S., Thibaud M.-C., Betsche T., Valadier M.-H. & Foyer C.H. (1997) Modulation of carbon and nitrogen metabolism, and of nitrate reductase, in untransformed Nicotiana plumbaginifolia during CO2 enrichment of plants grown in pots and hydroponic culture. Planta 202, 510–521.

Field C.B., Jackson R.B. & Mooney H.A. (1995) Stomatal responses to increased CO2: implications from plant to global scale. Plant, Cell and Environment 18, 1214–1225.

Furbank, R.T., W.C. Taylor. 1995. Regulation of photosynthesis in C3 and C4 plants: A molecular approach. Plant Cell, 7(7): 797-807. https://doi.org/10.1105/tpc.7.7.797.

George V., Gerant D. & Dizengremel P. (1996) Photosynthesis, Rubisco activity and mtichondrial malate oxidation in peduculate oak seedlings grown under present and elevated atmospheric carbon dioxide concentrations. Annales Des Sciences Forestrieres (Paris) 53, 469–474.

Gerhart LM, Ward JK. 2010. Plant responses to low [CO2] of the past.NewPhytologist188: 674–695.

Global Monitoring Laboratory (https://gml.noaa.gov/ccgg/trends/)

Gojon, A., Cassan, O., Bach, L., Lejay, L. and Martin, A., 2023. The decline of plant mineral nutrition under rising CO2: physiological and molecular aspects of a bad deal. Trends in Plant Science, 28(2), pp.185-198.

Gunderson C.A., Norby R.J. & Wullschleger S.D. (1993) Foliar gas exchange of two deciduous hardwoods during three years of growth in elevated CO2: no loss of photosynthetic enhancement. Plant, Cell and Environment 16, 797–807.

Hassan, M.R. and Ito, D., 2023. Down-regulation of photosynthesis in apple leaves under elevated CO2 concentration: a long-term field study with different fruit load. Journal of Agricultural Meteorology, 79(1), pp.49-57.

Heath J. & Kerstiens G. (1997) Effects of elevated CO2 on leaf gas exchange in beech and oak at two levels of nutrient supply: consequences for sensitivity to drought in beech. Plant, Cell and Environment 20, 57–67.

Hurry, V.M. and Huner, N.P., 1991. Low growth temperature effects a differential inhibition of photosynthesis in spring and winter wheat. Plant Physiology, 96(2), pp.491-497.

Hurry, V.M. and Huner, N.P., 1992. Effect of cold hardening on sensitivity of winter and spring wheat leaves to short-term photoinhibition and recovery of photosynthesis. Plant Physiology, 100(3), pp.1283-1290.

Idso S.B. & Kimball B.A. (1992) Effects of atmospheric CO2 enrichment on photosynthesis, respiration and growth of sour orange trees. Plant Physiology 99, 341–343.

IPCC (2021). "Summary for policymakers," in Climate change 2021: The physical science basis. contribution of working group I to the sixth assessment report of the intergovernmental panel on climate change. Eds. Masson-Delmotte, V., Zhai, P., Pirani, A., Connors, S. L., Péan, C., Berger, S., Caud, N., Chen, Y., Goldfarb, L., Gomis, M. I., Huang, M., Leitzell, K., Lonnoy, E., Matthews, J. B. R., Maycock, T. K., Waterfield, T., Yelekçi, O., Yu, R., Zhou, B. (Cambridge University Press).

IPCC (2023) Climate Change 2023: Synthesis Report. Contribution of Working Groups I, II and III to the Sixth Assessment Report of the Intergovernmental Panel on Climate Change. Geneva, Switzerland.

Jackson R.B., Sala O.E., Field C.B. & Mooney H.A. (1994) CO2 alters water use, carbon gain and yield in a natural grassland. Oecologia 98, 257–262.

Jacob J., Greitner C. & Drake B.G. (1995) Acclimation of photosynthesis in relation to Rubisco and non-structural carbohydrate contents and in situ carboxylase activity in Scirpus olneyi grown at elevated carbon dioxide in the field. Plant, Cell and Environment 18, 875–884.

Jang J.C. & Sheen J. (1994) Sugar sensing in higher plants. Plant Cell 6, 1665–1679.

Kaiser, E., Zhou, D., Heuvelink, E., Harbinson, J., Morales, A. and Marcelis, L.F., 2017. Elevated CO2 increases photosynthesis in fluctuating irradiance regardless of photosynthetic induction state. Journal of Experimental Botany, 68(20), pp.5629-5640.

Kant, S., Seneweera, S., Rodin, J., Materne, M., Burch, D., Rothstein, S.J. and Spangenberg, G., 2012. Improving yield potential in crops under elevated CO2: integrating the photosynthetic and nitrogen utilization efficiencies. Frontiers in Plant Science, 3, p.162.

Knapp A.K., Hamerlynck E.P. & Owensby C.E. (1993) Photosynthetic and water relations responses to elevated CO2 in the C4 grass Andropogon gerardii. International Journal of Plant Sciences 154, 459–466.

Kusano, M., Fukushima, A., Redestig, H. and Saito, K., 2011. Metabolomic approaches toward understanding nitrogen metabolism in plants. Journal of Experimental Botany, 62(4), pp.1439-1453.

Kutik J., Natr L., Demmers-Derks H.H. & Lawlor D.W. (1995) Chloroplast ultrastructure of sugar beet culitvated in normal and elevated carbon dioxide concentrations with two contrasted nitrogen supplies. Journal of Experimental Botany 46, 1797–1802.

Lammertsma EI, Jan De Boer H, Dekker SC, Dilcher DL, Lotter AF, Wagner-Cremer F. 2011. Global CO2rise leads to reduced maximum stomatalconductance in Florida vegetation. Proceedings of the National Academy of Sciences,USA108: 4035–4040.

Le Quere C, Andrew RM, Canadell JG, Sitch S, Korsbakken JI, Peters GP,Manning AC, Boden TA, Tans PP, Houghton RAetal.2016.Global carbonbudget 2016.Earth System Science Data8: 605–649.

Liang, J., Chen, X., Guo, P., Ren, H., Xie, Z., Zhang, Z. and Zhen, A., 2021. Grafting improves nitrogen-use efficiency by regulating the nitrogen uptake and metabolism under low-nitrate conditions in cucumber. Scientia Horticulturae, 289, p.110454.

Liu, W., Liu, L., Yan, R., Gao, J., Wu, S. and Liu, Y., 2023. A comprehensive meta-analysis of the impacts of intensified drought and elevated CO2 on forage growth. Journal of Environmental Management, 327, p.116885.

Long, M., Yunshanjiang, M., Yu, D., Li, S., Tuerdimaimaiti, M., Wu, A. and Zhang, G., 2023. Short-Term Elevated CO2 or O3 Reduces Undamaged Rice Kernels, but Together They Have No Effect. Agronomy, 13(12), p.2981.

Long, S.P., D.R. Ort. 2010. More than taking the heat: crops and global change. Curr. Opin. Plant Biol., 13(3): 241-248. https://doi.org/10.1016/j.pbi.2010.04.008.

Long SP, Ainsworth EA, Leakey ADB, N€osberger J, Ort DR. 2006.Food forthought: lower-than-expected crop yield stimulation with rising CO2concentrations.Science312: 1918–1921.

McCaig, T.N. and Hill, R.D., 1977. Cyanide-insensitive respiration in wheat: cultivar differences and effects of temperature, carbon dioxide, and oxygen. Canadian Journal of Botany, 55(5), pp.549-555.

McCarthy HR, Oren R, Johnsen KH, Gallet-Budynek A, Pritchard SG, Cook CW, LaDeau SL, Jackson RB, Finzi AC. 2010.Re-assessment of plant carbondynamics at the Duke free-air CO_2 enrichment site: interactions of atmospheric[CO_2] with nitrogen and water availability over stand development.NewPhytologist185: 514–528.

Mcconnaughay KDM, Berntson GM, Bazzaz FA. 1993. Limitations to CO_2-induced growth enhancement in pot studies.Oecologia94: 550–557.

Moore B.D., Cheng S.-H., Rice J. & Seeman J. (1998) Sucrose cycling, Rubisco expression and prediction of photosynthetic acclimation to elevated carbon dioxide. Plant, Cell and Environment 21, 905–915.

Moore BD, Cheng S-H, Sims D, Seemann JR. 1999.The biochemical andmolecular basis for photosynthetic acclimation to elevated atmospheric CO2.Plant, Cell & Environment22: 567–582.

Morison J.I.L. (1987) Intercellular CO2 concentration and stomatal response to CO2. In Stomatal Function (eds Z. Zeiger, G.D. Farquhar & I.R. Cowan), pp. 229–251. Stanford University Press, Stanford, CA.

National Centers for Environmental Information (https://www.ncei.noaa.gov/)

Nie G.Y., Hendrix D.L., Webber A.N., Kimball B.A. & Long S.P. (1995b) Increased accumulation of carbohydrates and decreased photosynthetic gene transcript levels in wheat grown at an elevated carbon dioxide concentration in the field. Plant Physiology 108, 975–983.

Obermeier WA, Lehnert LW, Kammann CI, M€uller C, Gr€unhage L, LuterbacherJ, Erbs M, Moser G, Seibert R, Yuan Netal.2017.Reduced CO2fertilizationeffect in temperate C3grasslands under more extreme weather conditions.NatureClimate Change7: 137–141.

Oechel E.C., Hastings J., Vourlitis M., Jenkins M., Riechers G. & Grulke N. (1993) Recent change of artic tundra ecosystems from a net carbon dioxide sink to a source. Nature 361, 520–523.

Oquist, G., Hurry, V.M. and Huner, N.P., 1993. Low-temperature effects on photosynthesis and correlation with freezing tolerance in spring and winter cultivars of wheat and rye. Plant Physiology, 101(1), pp.245-250.

Owensby C.E., Ham J.M., Knapp A.K., Bremer D. & Aven L.M. (1997) Water vapour fluxes and their impact under elevated CO2 in a C4-tallgrass prairie. Global Change Biology 3, 189–195.

Park, S., Park, B.R., Jeong, D., Park, J., Ko, J.K., Kim, S.J., Kim, J.S., Jin, Y.S. and Kim, S.R., 2023. Functional expression of RuBisCO reduces CO2 emission during fermentation by engineered Saccharomyces cerevisiae. Process Biochemistry, 134, pp.286-293.

Pedrotti, L., Weiste, C., Nägele, T., Wolf, E., Lorenzin, F., Dietrich, K., Mair, A., Weckwerth, W., Teige, M., Baena-González, E. and Dröge-Laser, W., 2018. Snf1-RELATED KINASE1-controlled C/S1-bZIP signaling activates alternative mitochondrial metabolic pathways to ensure plant survival in extended darkness. The Plant Cell, 30(2), pp.495-509.

Petereit, J., Katayama, K., Lorenz, C., Ewert, L., Schertl, P., Kitsche, A., Wada, H., Frentzen, M., Braun, H.P. and Eubel, H., 2017. Cardiolipin supports respiratory enzymes in plants in different ways. Frontiers in plant science, 8, p.72.

Polley HW, Johnson HB, Marino BD, Mayeux HS. 1993.Increase in C3plantwater-use efficiency and biomass over glacial to present CO2concentrations.Nature361:61–64.

Poorter H, Fiorani F, Pieruschka R, Wojciechowski T, van derPutten WH, KleyerM, Schurr U, Postma J. 2016.Pampered inside, pestered outside? Differencesand similarities between plants growing in controlled conditions and in the field.New Phytologist212: 838–855.

Posch, B.C., Kariyawasam, B.C., Bramley, H., Coast, O., Richards, R.A., Reynolds, M.P., Trethowan, R. and Atkin, O.K., 2019. Exploring high temperature responses of photosynthesis and respiration to improve heat tolerance in wheat. Journal of experimental botany, 70(19), pp.5051-5069.

Pritchard S.G., Peterson C.M., Prior S.A. & Rogers H.H. (1997) Elevated carbon dioxide differentially affects needle ultrastructire and phloem anatomy in Pinus palustris: interactions with soil resource availability. Plant, Cell and Environment 20, 461–471.

Qiu, J., Cai, C., Shen, M., Gu, X., Zheng, L., Sun, L., Teng, Y., Yu, H. and Zou, L., 2023. Responses of growth, yield and fruit quality of strawberry to elevated CO2, LED supplemental light, and their combination in autumn through spring greenhouse production.

Rakhmankulova, Z.F., Shuyskaya, E.V., Prokofieva, M.Y., Saidova, L.T. and Voronin, P.Y., 2023. Effect of Elevated CO 2 and Temperature on Plants with Different Type of Photosynthesis: Quinoa (C 3) and Amaranth (C4). Russian Journal of Plant Physiology, 70(6), p.117.

Riviere-Rolland H., Contard P. & Betsche T. (1996) Adaptation of pea to elevated atmospheric CO2: Rubisco, PEP carboxylase and chloroplast phosphate translocator at different levels of nitrogen and phosphate nutrition. Plant, Cell and Environment 19, 109–117.

Rogers A, Allen DJ, Davey PA, Morgan PB, Ainsworth EA, Bernacchi CJ,Cornig G, Dermody O, Dohleman FG, Heaton EAetal.2004.Leafphotosynthesis and carbohydrate dynamics of soybeans grown throughout theirlife-cycle under free-air carbon dioxide enrichment. Plant, Cell & Environment27: 449–458.

Rogers G.S., Milham P.J., Gillings M. & Conroy J.P. (1996b) Sink strength may be the key to growth and nitrogen response in Ndeficient wheat at elevated carbon dioxide. Australian Journal of Plant Physiology 23, 253–264

Sadok, W., Lopez, J.R. and Smith, K.P., 2021. Transpiration increases under high-temperature stress: Potential mechanisms, trade-offs and prospects for crop resilience in a warming world. Plant, Cell & Environment, 44(7), pp.2102-2116.

Sage R.F. (1994) Acclimation of photosynthesis to increasing atmospheric carbon dioxide; the gas exchange perspective. Photosynthesis Research 39, 351–368.

Sekhar, K.M., Reddy, K.S.R. and Reddy, A.R., 2023. Photosynthesis and carbon sequestration efficacy of Conocarpus erectus L.(Combretaceae) grown under elevated CO2 atmosphere. Plant Physiology Reports, pp.1-11.

Sharkey TD, Bernacchi CJ, Farquhar GD, Singsaas EL. 2007. Fittingphotosynthetic carbon dioxide response curves for C3 leaves. Plant, Cell & Environment 30: 1035–1040.

Sharkey TD, Stitt M, Heineke D, Gerhardt R, Raschke K, Heldt HW. 1986.Limitation of photosynthesis by carbon metabolism: II. O2-insensitive CO2uptake results from limitation of triose phosphate utilization. Plant Physiology 81:1123–1129

Sharma, N., Thakur, M., Suryakumar, P., Mukherjee, P., Raza, A., Prakash, C.S. and Anand, A., 2022. 'Breathing Out'under heat stress—respiratory control of crop yield under high temperature. Agronomy, 12(4), p.806.

Sicher R.C. & Bunce J.A. (1997) Relationship of o photosynthetic acclimation to changes of Rubisco activity in field grown winter wheat and barley during growth in elevated carbon dioxide. Photosynthesis Research 52, 27–38.

Sicher R.C. & Kremer D.F. (1994) Responses of Nicotiana tabacum to carbon dioxide enrichment at low-photon flux density. Physiologia Plantarum 92, 383–388

Sicher R.C., Kremer D.F. & Rodermel S.R. (1994) Photosynthetic acclimation to elevated carbon dioxide occurs in transformed tobacco with decreased ribulose-1,5-bisphosphate carboxylase/ oxygenase content. Plant Physiology 104, 409–415.

Siebers, M. H., Slattery, R. A., Yendrek, C. R., Locke, A. M., Drag, D., Ainsworth, E. A., … Ort, D. R. (2017). Simulated heat waves during maize reproductive stages alter reproductive growth but have no lasting effect when applied during vegetative stages. Agriculture, Ecosystems and Environment, 240, 162–170.

Siebers, M. H., Yendrek, C. R., Drag, D., Locke, A. M., Rios Acosta, L., Leakey, A. D. B.,Ort, D. R. (2015). Heat waves imposed during early pod development in soybean (Glycine max) cause significant yield loss despite a rapid recovery from oxidative stress. Global Change Biology, 1–13.

Slattery, R.A. and Ort, D.R., 2019. Carbon assimilation in crops at high temperatures. Plant, cell & environment, 42(10), pp.2750-2758.

Smith, N.G. and Dukes, J.S., 2017. Short-term acclimation to warmer temperatures accelerates leaf carbon exchange processes across plant types. Global change biology, 23(11), pp.4840-4853.

Smith NG, Dukes JS. 2013.Plant respiration and photosynthesis in global-scalemodels: incorporating acclimation to temperature and CO2.Global ChangeBiology19:45–63.

Smith SD, Charlet TN, Zitzer SF, Abella SR, Vanier CH, Huxman TE. 2014.Long-term response of a Mojave Desert winter annual plant community to awhole-ecosystem atmospheric CO2manipulation (FACE).GlobalChange Biology20: 879–892.

Stitt M. (1991) Rising CO2 levels and their potential significance for carbon flow in photosynthetic cells. Plant, Cell and Environment 14, 741–762.

Teskey R.O. (1995) A field study of the effects of elevated CO2 on carbon assimilation, stomatal conductance and leaf and branch growth of Pinus taeda trees. Plant, Cell and Environment 18, 565–573.

Tissue D.T., Thomas R.B. & Strain B.R. (1993) Long-term effects of elevated CO2 and nutrients on photosynthesis and Rubisco in loblolly pine seedlings. Plant, Cell and Environment 16, 859–865.

Tissue D.T., Thomas R.B., Strain B. & R. (1997) Atmospheric CO2 enrichment increases growth and photosynthesis of Pinus taeda, a 4 year experiment in the field. Plant, Cell and Environment 20, 1123–1134.

Van Aken, O. and Van Breusegem, F., 2015. Licensed to kill: mitochondria, chloroplasts, and cell death. Trends in Plant Science, 20(11), pp.754-766.

Van Oosten J.-J. & Besford R.T. (1995) Some relationships between the gas exchange, biochemistry and molecular biology of photosynthesis during leaf development of tomato plants after transfer to different carbon dioxide concentrations. Plant, Cell and Environment 18, 1253–1256.

Van Oosten J.-J. & Besford R.T. (1996) Acclimation of photosynthesis to elevated carbon dioxide through feedback regulation of gene expression: climate of opinion. Photosynthesis Research 48, 353–365.

Van Oosten J.-J., Wilkins D. & Besford R.T. (1994) Regulation of the expression of photosynthetic nuclear genes by high CO2 is mimicked by carbohydrates: a mechanism for the acclimation of photosynthesis to high CO2. Plant, Cell Environment 17, 913–923.

von Ghannoum O., Caemmerer S., Barlow E.W.R. & Conroy J.P. (1997) The effect of carbon dioxide enrichment on the growth, morphology and gas exchanges of a C-3 (Panicum laxum) and a C-4 (Panicum antidotale) grass. Australian Journal of Plant Physiology 24, 227–237

Vu J.V.C., Allen L.H. Jr, Boote K.J. & Bowes G. (1997) Effects of elevated carbon dioxide and temperature on photosynthesis and Rubisco in rice and soybean. Plant, Cell and Environment 20, 68–76.

Wang, K. and Dickinson, R.E., 2012. A review of global terrestrial evapotranspiration: Observation, modeling, climatology, and climatic variability. Reviews of Geophysics, 50(2).

Wang, P. C., DU, Y. Y., AN, G. Y., Zhou, Y., Miao, C. and Song, C. P., 2006, Analysis of global expression profiles of Arabidopsis genes under abscisic acid and H_2O_2 applications. J. Integr. Plant Biol., 48(1): 62-74.

Wang X, Lewis JD, Tissue DT, Seemann JR, Griffin KL. 2001.Effects of elevatedatmospheric CO2concentration on leaf dark respiration ofXanthium strumariumin light and in darkness.Proceedings of the National Academy of Sciences, USA98:2479–2484.

Way, D. A., & Yamori, W. (2014). Thermal acclimation of photosynthesis: On the importance of adjusting our definitions and accounting for thermal acclimation of respiration. Photosynthesis Research, 119, 89–100.

Welchen, E. and Gonzalez, D.H., 2016. Cytochrome c, a hub linking energy, redox, stress and signaling pathways in mitochondria and other cell compartments. Physiologia Plantarum, 157(3), pp.310-321.

Yamasaki, T., Yamakawa, T., Yamane, Y., Koike, H., Satoh, K. and Katoh, S., 2002. Temperature acclimation of photosynthesis and related changes in photosystem II electron transport in winter wheat. Plant Physiology, 128(3), pp.1087-1097.

Yamori, W., Hikosaka, K., & Way, D. A. (2014). Temperature response of photosynthesis in C_3, C_4, and CAM plants: Temperature acclimation and temperature adaptation. Photosynthesis Research, 119, 101–117.

Yang, X., Wang, D., Tao, Y., Shen, M., Ma, C., Cai, C., Song, L., Yin, B. and Zhu, C., 2023. Does elevated CO2 enhance the arsenic uptake by rice? Yes or maybe: Evidences from FACE experiments. Chemosphere, 327, p.138543.

Yang, Z., Jiang, Y., Qiu, R., Gong, X., Agathokleous, E., Hu, W. and Clothier, B., 2023. Heat stress decreased transpiration but increased evapotranspiration in gerbera. Frontiers in Plant Science, 14, p.1119076.

Yuan, L., Tang, L., Zhu, S., Hou, J., Chen, G., Liu, F., Liu, S. and Wang, C., 2017. Influence of heat stress on leaf morphology and nitrogen–carbohydrate metabolisms in two wucai (Brassica campestris L.) genotypes. Acta Societatis Botanicorum Poloniae, 86(2).

Zancani, M., Casolo, V., Petrussa, E., Peresson, C., Patui, S., Bertolini, A., De Col, V., Braidot, E., Boscutti, F. and Vianello, A., 2015. The permeability transition in plant mitochondria: the missing link. Frontiers in Plant Science, 6, p.1120.

Zheng, Y., Li, F., Hao, L., Yu, J., Guo, L., Zhou, H., Ma, C., Zhang, X. and Xu, M., 2019. Elevated CO2 concentration induces photosynthetic down-regulation with changes in leaf structure, non-structural carbohydrates and nitrogen content of soybean. BMC plant biology, 19(1), pp.1-18.

Ziska L.H., Weerakoon W., Namuco O.S. and Pamplona R. 1996. The influence of nitrogen on the elevated carbon dioxide response in field grown rice. Australian Journal of Plant Physiology 23, 45–52.

5

Mitigation and Adaptation Strategies for Climate Change and Pest Management

Gargi C., Akshaja Suresh, Vyshnavi Sunil and Aura Senson

Department of Entomology, College of Agriculture, Vellayani, KAU Thiruvananthapuram, Kerala

Abstract

Climate change is a serious problem which threatens agriculture around the world. Rising level of atmospheric CO_2, increase of temperature, and altered precipitation patterns affect insect pests, natural enemies and pollinators in several ways. This ultimately ends up as a threat to global food security by reducing agricultural productivity. Pro- active and scientific methods like, climate-smart agricultural techniques, can manage these insects, properly. Modified integrated pest management strategies and proper climate and pest population monitoring systems can keep a check on increasing pest status of insects. To conserve beneficial insects, modification of agricultural landscapes and balanced use of chemicals are essential. Climate change will require eco-friendly, adaptive management strategies to effectively, deal with changing status of insects.

Keywords: *Climate change, insect pests, beneficial insects, management*

Introduction

The world's food and nutritional security are under serious threat from climate change. Climate changes have significant effects on agriculture by affecting insect pests, their natural enemies, and pollinators as illustrated in Fig 1 (Eigenbrode and Adhikari, 2023). According to Benedict (2003), 13.6% of annual crop loss worldwide is attributed to insect pest attack. Crops, crop pests, natural enemies and pollinators are directly and indirectly impacted by erratic climate. The continuous and anticipated alterations in these insects vary across species, systems, and geographical areas, impeding the development of strong generalizations and forecasts. Moreover, the implementation of "climate-smart" agricultural methods, may have additional effects on ideal management of these insects.

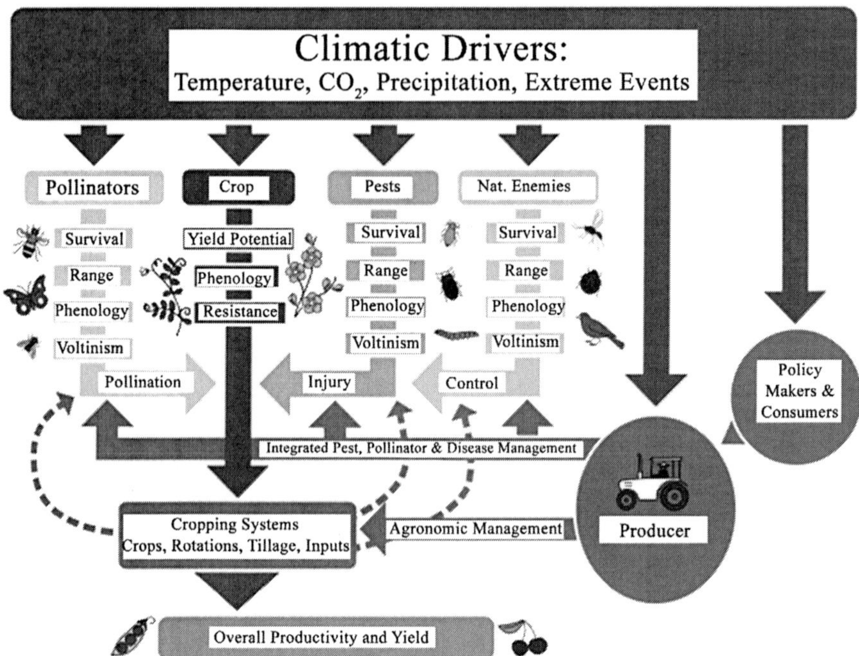

Fig. 1: The pathways through which climate change can affect crop pests and beneficial insects

Impact of climate change on insect pests

Direct effects on insect pests

Temperature: The most reliable, well-researched, and easily modelled component of climate change is warming. Warming affects physiology, phenology, dispersal, reproduction, development, and survival of insect pests (Bale *et al.*, 2002) as illustrated in Fig 2 (Skendžić *et al.*, 2021). It can be inferred that increased herbivory would be associated with rising temperatures in many cases. High temperature and high humidity are key reasons for whitefly population increase (Pathania *et al.*, 2020). Warming can alter the number of insect generations in *Helicoverpa armigera* (Lepidoptera: Noctuidae) in China (Huang and Hao, 2020).

Increased CO_2: Increased CO_2 can affect insect herbivores in a negative, positive, or neutral way, and the impacts can vary depending on the species (Eigenbrode and Adhikari, 2023). Consumption rates, growth rates, and multiplication of insect pests are affected by this increase (Fuhrer, 2003). C4 plants are less sensitive to increased CO_2 levels and are less susceptible to insect damage when compared with C3 plants (Lincoln *et al.*, 1984). In foliage feeders, increased consumption rates occur as response to a decrease in nitrogen, as anticipated by CO_2 fertilization (Skendžić *et al.*, 2021). Under elevated CO_2, survival of *Paropsis atomaria* (Coleoptera: Chrysomelidae) increased on *Eucalyptus robusta* but decreased on *Eucalyptus tereticornis* (Gherlenda *et al.*, 2015). Pupal weight and fecundity of artificial diet fed *Helicoverpa. armigera* were lower at 550 and 750 ppm CO_2 compared to that reared in ambient CO_2 (400 ppm) (Liu *et al.*, 2017).

Precipitation: As a part of changing climate, precipitation intensity has increased while precipitation frequency has decreased. This results in washing off of insect stages like eggs, larvae and also small-bodied pests. Rapid growth of wireworm populations in the upper part of the soil as a result of increased summer rainfall was recorded by Staley *et al.* 2007. Shifting aphid flight phenology is a result of changes in temperature and precipitation (Crossley *et al.*, 2022).

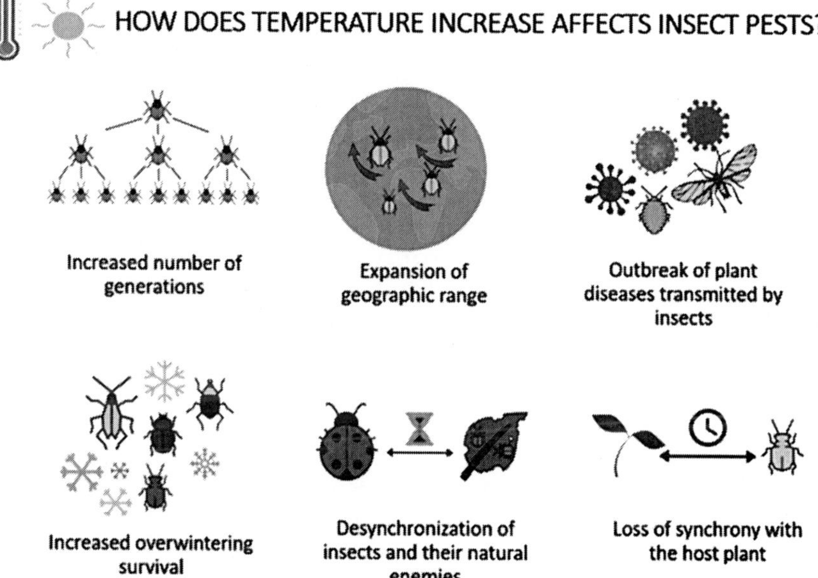

Fig. 2: Effects of temperature rise on agricultural insect pests

Effects through species interactions

The impacts of climate change on insect pests can be seen in their interactions with natural enemies with direct changes to their physiology or behaviour, disruption of their phenological overlap with pests, changes to their geographic ranges from those of their hosts, interactions involving alternative prey in agroecosystems (Eigenbrode and Adhikari, 2023). Slogget (2021) has reported that variations in aphid abundance could result in modifications to ladybird prey interactions and the tendency of ladybirds to feed on multiple aphid species variations based on their abundance could save ladybirds from decline.

Expansion of insects' distribution

The geographical dispersion of insect pests will be significantly impacted by climate change. Low temperatures frequently have a greater influence on this dispersal than high temperatures. By 2055, the ranges of insect pests are expected to expand to higher altitudes, with a rise in the number of generations in central Europe (Skendžić *et al.*, 2021).

There is direct correlation between invasive alien species introduction and climate change. Pests may be transported to new geographic locations by extreme weather events, where they may find favourable environmental circumstances for establishment. *Cactoblastis cactorum* (Lepidoptera: Pyralidae), was introduced to Mexico from Caribbean islands in 2005, by a hurricane (Skendžić *et al.*, 2021).

Effects on insecticides and host plant resistance

Warmer weather can reduce the effectiveness of some insecticides. Exposure to higher temperatures post exposure reduces the toxicity of bifenthrin, lambdacyhalothrin and spinosad to *Ostrinia nubilalis* (Lepidoptera: Crambidae) (Musser and Shelton, 2005). Decreased effectiveness of triazophos against the pest *Nilaparvata lugens* (Hemiptera: Delphacidae) in the presence of increased CO_2 has been documented (Ge *et al.*, 2013). Host plant resistance to pests can also be changed by higher temperatures as seen in most wheat genes imparting resistance to Hessian fly (*Mayetiola destructor*) (Diptera: Cecidomyiidae) which are sensitive to temperature (Tang *et al.*, 2018).

Impact of climate change on beneficial insects- natural enemies and pollinators

Effects of climate change on pollinators

Plant pollinator interactions

Climate change affect plants by reducing the insect visit and pollen deposition whereas it affect pollinators by reducing their food availability (Hegland *et al.*, 2009). In Japan, early flowering resulted in climate change causing early blooming of flowers which resulted in a lack of food for butterflies and other pollinators. Studies were conducted in Spain during 1952 and 2014 which revealed that *Apis mellifera* adapted quickly to the change in temperature and appeared early in the spring (Gordo and Sanz, 2006). Nectar and nutritional quality of pollen are effected by temperature changes, but pollen composition is more affected (Sato *et al.,* 2006). With increase in the temperature plants shifted their phenology. Also, insects like butterflies, beetles and moths also shifted their phenology in response to the change in climate (Freimuth *et al.,* 1971).

Changes in pollinator behaviour

A change in sensory cues due to climate change can affect the learning ability of pollinators (Chittka and Raine, 2006). With increase in temperature there will be changes in endogenous floral scent production that will affect the pollinator behaviour. They will find it difficult to detect the flower most likely insects like moths that depend on long distance cues to locate the flower (Kevan and Baker, 1983; Yuan *et al*., 2009).

Effect on plant-pollinator networks

Plants need pollinators for their propagation (Ollerton *et al.,* 2011) whereas pollinators require plants for pollen and nectar (Memmott *et al.,* 2007), Because of this interdependency change in population of either of these affect each other. Early emergence of bumblebee queens from hibernation were also found as a result of warmer soil temperature due to climate change (Alford, 1969)

Climate-driven shifts in phenology

Change in temperature can lead to a shift in phenology of plant causing asynchrony between pollinators and their food availability leading to less pollinator visitation in plants because plants and pollinators are coevolved to synchronise their activities (Donnelly *et al*., 2011). Modelling studies indicate that among 1420 pollinators and 429 plant species over last 120 years indicate phenological shift and reduced interaction between pollinators and their host plants (Burkle *et al*., 2013). But generalist pollinators can cope up with this by switching between host plants.

Effects of climate change on natural enemies

Effect on natural enemy distribution`

Harmonia axyridis (Coleoptera: Coccinellidae) is a coccinellid generalist predator with wide capacity for dispersal. They have colonized drier regions of Brazil too, but there was a decline in the population when temperature decreased in hot dry summers (Grez *et al*., 2017) which suggests the effect of climate change on distribution. A change in elevational distribution of *H. axyridis* due to climate change led to greater availability of preys (Roy *et al*.,

2016) and less exposure to unfavourable climatic conditions resulted in an increase in population.

Climate- driven shifts in ranges

Species that are unable to migrate cannot cope up with the changing climate, but migratory insects can easily cope up with this. Spatial and temporal changes in insect species can't be always linked to climate change because population dynamics are affected by other factors too for instance change in the land use pattern. Buff-tailed bumblebee (*Bombus terrestris*) and other generalist pollinators have been found to lose range in Southern Europe but gained some range in Northern Europe (Rasmont *et al.*, 2015). Losses from Southern range and shift to higher elevation by southern species have caused range contractions leading bumble bees unable to track their northern range (Settele *et al.*, 2016).

Host– enemy synchrony

Phenology of an insect is greatly affected by the temperature. An increase in temperature can cause early emergence of parasitoids resulting in an asynchrony between herbivore and parasitoid leading to starvation of herbivores which can result in their extinction. Many parasitoids die even before the availability of host. When hosts are unavailable, hibernating parasitiods of leaf miners *Cameraria ohridella* emerge resulting in a low level of parasitism of horse chestnut leafminers (Grabenweger *et al.*, 2007). To learn the effect of climate change on biocontrol, models are to be constructed.

Thermal and desiccation responses

Establishment of invasive predators depend upon the resistance of diapause to thermal extremes. If they enter into a diapause which is resistant to thermal extremes, they can easily establish. For instance, a coccinellid predator *H. axyridis* failed to establish in Azores when the factors responsible for inducing diapause were hindered (Soares *et al.*, 2008). Effect on natural enemies on host depend on decreased dry spell and temperature. A decreased parasitism on *Trichograma* eggs by European corn borer was observed when climate data was correlated with egg parasitism (Cagan *et al.*, 1998).

Management mediated changes

Due to poor plant growth, development time of herbivores is extended which increase their exposure to natural enemy decreasing their population. Drought tolerant cassava are more susceptible to mealybugs but show a decreased parasitism resulting in a reduction in pest control. Changing agricultural practices can also affect the natural enemy population. Mulching increased the population of parasitic diptera, predatory carabids, parasitoid hymenoptera and spiders in vineyard (Thomson and Hoffmann, 2007). Lands with crop to non-crop mosaics are best for conservation biocontrol (Tscharntke *et al.*, 2007). Abundance of predators like carabids differ between irrigated and non-irrigated fields whereas population of beetles and spiders are negatively affected by the droughts (Frampton *et al.*, 2000; Morecroft *et al.*, 2002).

Management of impact of climate change on beneficial insects and pests.

Climate change has exerted a pressure to modify the growth techniques to maximize, increase, or sustain productivity as it has disturbed the normal plant physiology and phenology. Use of agrochemicals in excessive amount to counteract the rise in harmful insect pests and diseases caused by changes in climatic conditions has negatively impacted the utilizing of arthropods for pollination and biological control (Sloggett, 2021). Pollinators and the services they provide can be directly or indirectly impacted by climate change very frequently. This happens when anticipated warming trends lead the timing of pollinator activity and flowering to change and become out of synchronisation, impacting both integrated pest management strategies and the services that pollinators perform. (Freimuth *et al.*, 2022).

Some techniques that can be integrated into the present practices of agriculture which can mitigate the effect of climate change on insect populations and their interactions with agriculture are given below.

Knowing the impact of climate, its change and impact on agriculture (Climate- smart farming).

Farming methods can be modified to adapt to and alleviate the consequences of climate change on the natural enemies, pollinators, and pests. The impact of climate change on agricultural output is fuelling the movement to alter the food system. Whether applied at the field, farm, or landscape scale, climate-resilient or "climate-smart" agriculture aims to boost agricultural profitability and productivity while enhancing resilience to climate change, lowering greenhouse gas emissions, and increasing carbon sequestration in farmland soils through a variety of techniques (Dinesh *et al.*, 2021).

Crop diversification and integrated cropping systems

Natural enemies and pests: There has long been interest in how cropping system variety affects insect pests and their natural enemies (Root, 1973). Hundreds of researches have examined the impacts of crop and weed diversity on both beneficial insects and pests, and these findings have been evaluated on a regular basis. Reduced pest pressure is typically associated with diversification, while there is clear variation in both intercropping and cover crops. Theoretically, less pests with crop diversification can arise from either reduced host finding abilities and pest population growth (the Resource Concentration Hypothesis), enhanced generalist natural foes (the Natural foes Hypothesis), or both. While there is some evidence to support each of these theories, the mechanisms underlying the reduction or increase in pest pressure that come with diversification are sometimes unclear or the product of several interrelated ecological factors. Additionally, diversification may make pest issues worse. For instance, by housing pests that must be controlled within the cover crop or by harboring pests that have the potential to harm a nearby or future cash crop, cover crops can increase the burden on pests. So the selection of crops in the system is vital. Insect vectors have the ability to transfer insect-borne plant infections that infect a main crop and a cover crop, hence raising the risk of infection and damage (Wegulo *et al.*, 2008).

Pollinators: By supplying more nectar and pollen sources, diversification involving the addition of flowering species encourages the growth of more pollinator species and maintains their numbers. Pollinator variety and abundance are increased by intercropping, also known as multiple cropping, in which forbs are planted alongside cereals (for example, legumes with wheat). Depending on the crop species included in the cover crop mix, flowering cover crops may also assist pollinators. Diversification of crop systems can counteract the detrimental impacts of agricultural intensification on pollinators (Brandmeier *et al.*, 2021).

Reduction in disturbances to soil

Natural enemies and pests: Tillage reduction increases the biodiversity of agroecosystems and the related ecosystem benefits. According to a meta-analysis comprising 59 studies, there was no significant difference in the total abundance of insect pests between reduced and conventional tillage. However, foliar pests were more prevalent and soil-dependent predators were less prevalent under high-disturbance tillage as opposed to low-disturbance tillage. Because tillage removes alternative food supplies and disturbs their habitats, it tends to have a negative impact on the abundance, species richness, and variety of some ground dwelling predators, particularly ground beetles (Coleoptera: Carabidae). However, tillage's effects on ground beetles vary, perhaps as a result of how it interacts with other management techniques. Furthermore, whereas plots with less tillage had fewer carabids in a long-term trial with spring cereals, there was no effect on another important ground predatory taxon, rove beetles (Coleoptera: Staphylinidae) (Andersen, 2003).

Pollinators: Over 70% of species of wild bees nest in the soil, where they spend a large portion of their life cycle. These ground-nesting bees are disturbed by tillage of any severity (Harmon-Threatt, 2020). Thus, conservation, decreased, or zero tillage can help pollinators and crop pollination in addition to improving crop productivity, farm profitability, and resistance to disturbances.

Reduction in chemical use on ecosystem

Beneficial insects and pests: Extended reliance on insecticides for pest management has been linked to well-documented detrimental effects on non-target organisms, such as natural enemies of pests that can spark secondary pest outbreaks, raise production costs, pose risks to human and environmental health, and make target pests more resistant to pesticides (Devi *et al.,* 2022). Decades of research and application have been driven by the premise of reducing pesticide use to overcome these vulnerabilities. Those who pollinate: additionally, numerous studies have demonstrated the harmful effects of fungicides, herbicides, and insecticides on pollinators, both directly and indirectly. As part of climate-smart agriculture, lowering these inputs will contribute to the preservation of ecosystem services including pollination and biological control (Bloom *et al.*, 2021).

Usage of amendments

Research on biochar and insect pests in agricultural systems is confined to how biochar affects soil degradation and pesticide persistence. It is feasible that biochar will change insect behaviour indirectly by affecting agricultural plants (Varjani *et al.*, 2019).

Modification of agricultural landscapes for changing climate

Pests and natural enemies: Landscape diversification can influence pests, especially by providing habitat for natural enemies. This can include woodlots, hedgerows, and semi natural habitats on the farm or landscape scale. However, a review of a large number of studies has shown that the benefits of pest control can vary, sometimes with trade-offs that may lead to reduced yields, although there are recognised principles for the management of agricultural landscapes for beneficial use (Karp *et al.*, 2018). The diversity of farms and landscapes also affects pollinating insects. Climate smart practices, such as increased diversity of agricultural landscape with seminatural habitats, field boundaries, beetles banks and hedgerows, native flower strips are enhancing pollinator biodiversity. For example, hedgerows support pollinators in adjacent orchards and vegetable fields, even weeds, which are present in semi natural habitats and field borders, provide pollinators with flowers that help them survive when there are no other resources available (Donkersley *et al.*, 2023).

Conclusion

Increased temperature, increased atmospheric CO_2, altered precipitation patterns, and other small-scale climate variability are some of the various factors of climate change that impact insect life. Different insect species respond differently to global warming depending on species heterogeneity, their hosts, and the climate variability. The impacts of climate change on insects are multifaceted, as it favours certain insects while hindering others. The major problems associated with climate change on insect pests are expansion of geographical location, development of multiple generations, increase in insect transmitted plant diseases and reduced effectiveness of management various management strategies. Climate change also impact the services offered by beneficial insects by altering their life cycle and host preferences. There will be considerable dent in global economy and huge threat to food security, if climate change alters normal insect life. To address this issue, proactive and scientific strategies like planning and developing improved IPM techniques, pest and climate monitoring and application of modelling tools are essential for managing pest attack and to conserve beneficial insects modification of agricultural landscapes and balanced use of chemicals, are imperative.

References

Alford, D.V. 1969. A study of hibernation of bumblebees (Hymenoptera-Bombidae) in Southern England. J. Anim. Ecol., 38, pp. 149–170.

Andersen, A. 2003. Long-term experiments with reduced tillage in spring cereals. II. Effects on pests and beneficial insects. Crop Protection, 22, pp. 147–152.

Bale, J.S., Masters, G.J., Hodkinson, I.D., Awmack, C., Bezemer, T.M., Brown, V.K., Butterfield, J., Buse, A., Coulson, J.C., Farrar, J. and Good, J.E., 2002. Herbivory in global climate change research: direct effects of rising temperature on insect herbivores. Global change biology, 8(1), pp.1-16.

Benedict, J.H. 2003. Strategies for controlling insect, mite and nematode pests. In: Plants, Genes and Crop Bio-technology, M.J. Chrispeels and D.E. Sadava (eds). Jones and Bartlet Publishers, Sudbury, MA, USA, pp. 414-442.

Bloom, E. H., Wood, T. J., Hung, K.-L. J., Ternest, J. J., Ingwell, L. L., Goodell, K., Kaplan, I., and Szendrei, Z. 2021. Synergism between local- and landscape-level pesticides reduces wild bee floral visitation in pollinator-dependent crops. Journal of Applied Ecology, 58, pp. 1187– 1198.

Brandmeier, J., Reininghaus, H., Pappagallo, S., Karley, A. J., Kiær, L. P. and Scherber, C., 2021. Intercropping in high input agriculture supports arthropod diversity without risking significant yield losses. Basic and Applied Ecology, 53, pp. 26–38.

Burkle, L. A., Marlin, J. C., and Knight, T. M., 2013. Plant-pollinator interactions over 120 years: loss of species, co-occurrence, and function, Science .339, 1611–1615.

Cagan, L., Tancik, J. and Hassan, S.A., 1998. Natural parasitism of European corn borer eggs Ostrinia nubilalis Hbn. (Lep., Pyralidae) by Trichogramma in Slovakia-need for field releases of the natural enemy. Journal of Applied Entomology. 122, pp. 315–318.

Chittka L. and Raine N.E. 2006. Recognition of flowers by pollinators. Curr. Opin. Plant Biol. 9, pp. 428–435

Crossley, M.S., Lagos-Kutz, D., Davis, T.S., Eigenbrode, S.D., Hartman, G.L., Voegtlin, D.J. and Snyder, W.E., 2022. Precipitation change accentuates or reverses temperature effects on aphid dispersal. Ecological Applications, 32(5), p.e2593.

Devi, P. I., Manjula, M. and Bhavani, R. V., 2022. Agrochemicals, environment, and human health. Annual Review of Environment and Resources, 47, pp. 399–421.

Dinesh, D., Hegger, D. L. T., Klerkx, L., Vervoort, J., Campbell, B. M. and Driessen, P. P. J., 2021. Enacting theories of change for food systems transformation under climate change. Global Food Security, 31, 100583.

Donkersley, P., Witchalls, S., Bloom, E. H., and Crowder, D. W. 2023. A little does a lot: Can small-scale planting for pollinators make a difference? Agriculture, Ecosystems & Environment, 343,p. 108254

Donnelly, Alison, Amelia Caffarra, and Bridget F. O'Neill., 2011. A review of climate-driven mismatches between interdependent phenophases in terrestrial and aquatic ecosystems. International Journal of Biometeorology, 55, pp. 805-817.

Eigenbrode, S.D. and Adhikari, S., 2023. Climate change and managing insect pests and beneficials in agricultural systems. Agronomy Journal, 115(5), pp.2194-2215.

Frampton, G. K., van den Brink, P. J. and Gould, P. J. L., 2000. Effects of spring drought and irrigation on farmland arthropods in southern Britain. Journal of Applied Ecology. 37, pp. 865–883.

Freimuth J, Bossdorf O, Scheepens JF. and Willems FM., 2022. Climate warming changes synchrony of plants and pollinators. Proceedings of the Royal Society B. Biological Sciences 289:20212142.

Fuhrer, J., 2003. Agroecosystem responses to combinations of elevated CO_2, ozone, and global climate change. Agriculture, Ecosystems & Environment, 97(1-3), pp.1-20.

Ge, L.Q., Wu, J.C., Sun, Y.C., Ouyang, F. and Ge, F., 2013. Effects of triazophos on biochemical substances of transgenic Bt rice and its nontarget pest Nilaparvata lugens Stål under elevated $CO_{,2}$. Pesticide biochemistry and physiology, 107(2), pp.188-199.

Gherlenda, A.N., Haigh, A.M., Moore, B.D., Johnson, S.N. and Riegler, M., 2015. Responses of leaf beetle larvae to elevated CO_2 and temperature depend on Eucalyptus species. Oecologia, 177, pp.607-617.

Gordo O and Sanz J. 2006.Temporal trends in phenology of the honey bee Apis mellifera (L.) and the small white Pieris rapae (L.) in the Iberian peninsula., Ecological Entomology.; 31, pp. 261– 268.

Grabenweger, G., Hopp, H., Jackel, B., Balder, H., Koch, T., Schmolling, S., 2007. Impact of poor host-parasitoid synchronisation on the parasitism of Cameraria ohridella (Lepidoptera: Gracillariidae). European Journal of Entomology. 104, pp. 153–158

Grez ,A. A., Zaviezo, T., Roy, H. E., Brown, P. M. J., Bizama, G. 2017. Rapid spread of Harmonia axyridis in Chile and its effects on local coccinellid biodiversity. Divers Distrib 22, pp. 982–994

Harmon-Threatt, A. 2020. Influence of nesting characteristics on health of wild bee communities. Annual Review of Entomology, 65, pp. 39–56.

Hegland, S. J., Nielsen, A., Lázaro, A., Bjerknes, A. L. and Totland, Ø., 2009. How does climate warming affect plant-pollinator interactions? Ecol Letters, 12, pp. 184-195.

Huang, J. and Hao, H., 2020. Effects of climate change and crop planting structure on the abundance of cotton bollworm, Helicoverpa armigera (Hübner) (Lepidoptera: Noctuidae). Ecology and evolution, 10(3), pp.1324-1338.

Karp, D. S., Chaplin-Kramer, R., Meehan, T. D., Martin, E. A., Declerck, F., Grab, H., Gratton, C., Hunt, L., Larsen, A. E., Martinez-Salinas, A., O'rourke, M. E., Rusch, A., Poveda,

K., Jonsson, M., Rosenheim, J. A., Schellhorn, N. A., Tscharntke, T., Wratten, S. D., Zhang, W., . . . Zou, Y. (2018). Crop pests and predators exhibit inconsistent responses to surrounding landscape composition. Proceedings of the National Academy of Sciences, 115, E7863–E7870.

Kevan, P. G, Baker, H. G. Insects as flower visitors and pollinators.1983. Annual Review of Entomology,28, pp. 407–453.

Lincoln, D.E., Sionit, N. and Strain, B.R., 1984. Growth and feeding response of Pseudoplusia includens (Lepidoptera: Noctuidae) to host plants grown in controlled carbon dioxide atmospheres. Environmental entomology, 13(6), pp.1527-1530.

Liu, J., Huang, W., Chi, H., Wang, C., Hua, H. and Wu, G., 2017. Effects of elevated CO_2 on the fitness and potential population damage of Helicoverpa armigera based on two-sex life table. Scientific Reports, 7(1), p.1119.

Memmott, J., Craze, P., Waser, N, and Price, M., 2007. Global warming and the disruption of plantpollinator interactions. Ecology Letters, 10. pp.710–717.

Morecroft, M. D., Bealey, M. D., Howells, O., Rennie, S. and Woiwod, I. P., 2002. Effects of drought on contrasting insect and plant species in the UK in the mid-1990s. Global Ecology and Biogeography. 11, pp. 7–22.

Musser, F.R. and Shelton, A.M., 2005. The influence of post-exposure temperature on the toxicity of insecticides to Ostrinia nubilalis (Lepidoptera: Crambidae). Pest Management Science: Formerly Pesticide Science, 61(5), pp.508-510.

Ollerton, J. 2011. How many flowering plants are pollinated by animals? Oikos, 120, pp. 321 326.

Pathania, M., Verma, A., Singh, M., Arora, P.K. and Kaur, N., 2020. Influence of abiotic factors on the infestation dynamics of whitefly, Bemisia tabaci (Gennadius 1889) in cotton and its management strategies in North-Western India. International journal of tropical insect science, 40, pp.969-981.

Rasmont, P., Franzen, M., Lecocq, T., Harpke, A., Roberts, S.P., Biesmeijer, J.C., Castro, L., Cederberg, B., Dvorak, L., Fitzpatrick, U. and Gonseth, Y., 2015. Climatic risk and distribution atlas of European bumblebees, 10, pp. 1-236. Pensoft Publishers.

Root, R. B. 1973. Organization of a plant-arthropod association in simple and diverse habitats—Fauna of collards (Brassica oleracea). Ecological Monographs, 43, pp. 95–120.

Roy, H. E., Brown, P. M. J., Adriaens, T., Berkvens, N., Borges, I., Clusella-Trullas, S., Comont ,R. F., De Clercq, P., Eschen, R., Estoup, A., Evans, E. W., Facon, B., Gardiner, M. M., Gil, A., Grez, A. A., Guillemaud, T., Haelewaters, D., Herz, A., Honek, A., Howe, A. G., Hui, C., Hutchison, W. D., Kenis, M., Koch, R. L., Kulfan, J., Lawson Handley, L., Lombaert, E., Loomans, A., Losey, J., Lukashuk, A. O., Maes, D., Magro, A., Murray, K. M., San Martin, G., Martinkova, Z., Minnaar, I. A., Nedved, O., Orlova-Bienkowskaja, M. J., Osawa, N., Rabitsch, W., Ravn, H. P., Rondoni, G., Rorke, S. L., Sergey, K., Ryndevich, S. K., Saethre, M. G., Sloggett, J. J, Soares, A. O., Stals, R., Tinsley, M. C., Vandereycken, A., van Wielink, P., Vigla´sˇova´, S., Zach, P., Zakharov, I. A., Zaviezo, T., Zhao, Z. 2016. The harlequin ladybird: global perspectives on invasion history and ecology. Biol Invasions 18: pp. 997–1044.

Sato, S., Kamiyama, M., Iwata, T., Makita, N., Furukawa H. and Ikeda, H., 2006. Moderate increase of mean daily temperature adversely affects fruit set of Lycopersicon esculentum by disrupting specific physiological processes in male reproductive development. Ann. Bot. 97, pp. 731–738.

Settele, J., Bishop, J., and Potts, S. G. 2016. Climate change impacts on pollination. Nature Plants, 2(7), pp.1-3.

Skendžić, S., Zovko, M., Živković, I.P., Lešić, V. and Lemić, D., 2021. The impact of climate change on agricultural insect pests. Insects, 12(5), p.440.

Sloggett, J.J., 2021. Aphidophagous ladybirds (Coleoptera: Coccinellidae) and climate change: a review. Insect Conservation and Diversity, 14(6), pp.709-722.

Soares, A.O., Borges, I., Borges, P.A.V., Labrie, G. and Lucas, E., 2008. Harmonia axyridis: What will stop the invader? Biocontrol 53, pp. 127–145.

Staley, J.T., Hodgson, C.J., Mortimer, S.R., Morecroft, M.D., Masters, G.J., Brown, V.K. and Taylor, M.E., 2007. Effects of summer rainfall manipulations on the abundance and vertical distribution of herbivorous soil macro-invertebrates. European Journal of Soil Biology, 43(3), pp.189-198.

Tang, G., Liu, X., Chen, G.H., Witworth, R.J. and Chen, M.S., 2018. Increasing temperature reduces wheat resistance mediated by major resistance genes to Mayetiola destructor (Diptera: Cecidomyiidae). Journal of economic entomology, 111(3), pp.1433-1438.

Thomson, L.J., Robinson, M. and Hoffmann, A.A., 2001., Field and laboratory evidence for acclimation without costs in an egg parasitoid. Functional Ecology. 15, 217–221

Tscharntke, T., Bommarco, R., Clough, Y., Crist, T.O., Kleijn, D., Rand, T.A., Tylianakis, J.M., van Nouhuys, S., Vidal, S., 2007. Conservation biological control and enemy diversity on a landscape scale. Biological Control, 43, 294–309

Varjani, S., Kumar, G., and Rene, E. R. 2019. Developments in biochar application for pesticide remediation: Current knowledge and future research directions. Journal of Environmental Management, 232,pp. 505–513.

Wegulo, S. N., Hein, G. L., Klein, R. N. and French, R. C. 2008. Managing wheat streak mosaic. University of Nebraska Lincoln. https://extensionpubs.unl.edu/publication/9000016365448/ managing–wheat–streak–mosaic/

Yuan, J.S., Himanen, S.J., Holopainen, J.K., Chen, F. and Stewart, C.N., 2009. Smelling global climate change: mitigation of function for plant volatile organic compounds. Trends in Ecology & Evolution, 24(6), pp.323-331.

6

Mitigation and Adaptation Strategies for Climate Change and Plant Diseases Management

Alby John, Saru Sara Sam and Amrith Raj

College of Agriculture, Vellayani, Thiruvananthapuram, Kerala

Abstract

Plant diseases are major problem for food production including quality and safety of food. Climate change has significant implications for plant pathology. Changes in temperature, precipitation patterns, humidity, and extreme weather events associated with climate change can directly and indirectly affect plant health and the dynamics of plant diseases. Crop mainly depends on climate and more than 50 percent of variation in the yield of a crop is due to climate differences. Climate change directly affects rate of physiological and biochemical processes of plants. These changes due to altered climatic patterns interferes in plant-pathogen interaction, shifts in disease distribution, altered disease dynamics, increased pathogen diversity, impact on plants defence mechanism and the influence on development and distribution of disease vectors. Climate change is expected to have profound effects on plant pathology, with implications for agriculture, forestry, natural ecosystems, and global food security.

Keywords: *Climate change, disease triangle, plant-pathogen interaction, disease severity, defence mechanism.*

Introduction

The “plant disease triangle” is a fundamental concept in plant pathology that illustrates the interaction between three essential components necessary for the development of a plant disease. Firstly, the host plant which refers to the plant species or cultivar that is susceptible to the pathogen. Host plants provide the necessary environment and resources for the pathogen to establish, grow, and reproduce. Secondly, the pathogen which is any organism (such as fungi, bacteria, viruses, nematodes, or other microorganisms) that can cause disease in plants. Pathogens can infect host plants through various means, including direct contact, airborne spores, soil-borne organisms, or vectors such as insects. And thirdly, the environment which encompasses all external factors that influence the interaction between the host plant and the pathogen. This includes abiotic factors such as temperature, humidity, soil moisture, light, and nutrient availability, as well as biotic factors such as the presence of other organisms (competitors, predators, symbionts) that may affect disease development.

The plant disease triangle illustrates that the occurrence of a disease requires the presence and interaction of all three components: a susceptible host, a virulent pathogen, and a conducive environment. If any of these components is missing or unfavourable, disease development may be inhibited or prevented. Any change in ecosystem can affect plant diseases, because plant disease is the result of interaction between a susceptible host, virulent pathogen and favourable environment. If any of the three factors are altered, the progress of any one disease can change. Although interactions among the three factors must be matched, weather is an important variable due to its dynamic behaviour.

Major parameters of climate change on plant diseases

1. Effect of moisture

Moisture can manifest as dew, rain, or irrigation water on plant surfaces (known as leaf wetness) and as relative humidity in the air. The presence of moisture is essential for spore germination and the penetration of fungi into host tissues. It also activates bacterial, fungal, and nematode pathogens prior to infecting the host. Moisture promotes sporulation, spore release, and germination in many fungi, bacterial multiplication and oozing on host surfaces, and facilitates nematode movement. High moisture levels enable these processes to occur continuously, leading to epidemics. Splashing rain and running water significantly contribute to the distribution and spread of pathogens.

Extended periods of high moisture, whether from rain, dew, or high relative humidity, are the primary factors driving epidemics caused by fungi (late blights, downy mildews, leaf spots, root rots, soft rots, rusts, and anthracnose), bacteria (leaf spots, blights, and soft rots), and nematodes. Devastating plant pathogens such as *Magnaporthe oryzae* (rice blast fungus) or *Puccinia striiformis* (stripe rust fungus) require a minimum of 5 hours of leaf wetness for disease to occur (Magarey *et al.*, 2005). As air can hold more water vapor at higher temperatures, the possibility for dew accumulation increases at higher temperatures. High moisture also increases the succulence of host plants, making them more susceptible to infection. The incidence of many diseases in specific regions correlates closely with the amount and distribution of rainfall throughout the year. Diseases like late blight of potatoes, downy mildew of grapes and cucurbits, bud rot of coconut, abnormal leaf fall of rubber, and foot rot in black pepper are more severe in areas and times with high rainfall or relative humidity.

The frequency of infection cycles for many fungal diseases corresponds with the number of rainy days per season. For soil-borne diseases causing root rots, damping-offs, seedling blights, and collar rots, disease severity increases with soil moisture, particularly near saturation points. Wet soil promotes the multiplication and movement of zoospores but reduces the host's ability to defend itself due to decreased oxygen availability in waterlogged soils. Soil-borne fungi like *Phytophthora*, *Pythium*, *Rhizoctonia*, *Fusarium*, *Sclerotinia*, and *Sclerotium*, as well as bacteria such as *Erwinia* and *Pseudomonas*, and most nematodes, cause the most damage in plants with wet but not flooded soils. Lower soil moisture decreases the incidence of infection of *Ralstonia solanacearum* in tomato plants (Islam and Toyota, 2004).

Certain pathogens, such as *Fusarium solani* (dry root rots), *F. roseum* (seedling blight), *Macrophomina phaseoli* (charcoal rot), and *Streptomyces scabies* (common scab of potatoes), are more severe in dry weather on plants stressed by water deficiency and seldom develop into epidemics (Johansen *et al.*, 2015). Fungi causing powdery mildew diseases can germinate, penetrate, and infect hosts without free water in dry conditions. In areas experiencing decreasing moisture due to climate change, these pathogen populations may increase. Diseases caused by viruses and phytoplasmas are indirectly influenced by moisture through its effects on vector activity. Moisture enhances the activity of fungal and nematode vectors of viruses while reducing the activity of aphids, plant hoppers, and mealybugs.

2. Effect of temperature

Plants and pathogens have specific minimum, maximum, and optimum temperature requirements for growth and survival. For example, a temperature of 15°C is ideal for *Globodera pallida* nematodes to infect potato plants (Jones *et al.*, 2017). Daytime temperatures of 35°C and night temperatures of 27°C are optimal for *Xanthomonas oryzae* bacteria to thrive in rice (Horino *et al.*, 1982). Meanwhile, temperatures ranging from 26°C to 31°C are best suited for the papaya ringspot virus (PRSV) to infect papaya (Mangrauthia *et al.*, 2009). In temperate regions, the low temperatures of winter are generally unfavourable for most pathogens. However, some fungi thrive at lower temperatures, and variations can exist among races of the same pathogen. Temperature also influences the production and release of spores, with many diseases thriving in cooler temperatures while others prefer higher temperatures. Diseases like fusarium wilts, anthracnose, rusts, powdery mildews, and bacterial wilts of solanaceous crops are more prevalent in hot tropical and subtropical regions. For instance, the late blight pathogen of potatoes (*Phytophthora infestans*) is most severe in northern latitudes, becoming a concern in subtropical regions only during winter.

Warming temperatures may also favour the emergence of new pathogen strains adapted to these conditions. For instance, the response to temperature varies among different populations of the *P. infestans* pathogen (Mariette *et al.*, 2016). Since 2000, more aggressive strains of the rust fungus *Puccinia striiformis* affecting wheat have become dominant worldwide, especially at higher temperatures. However, not all diseases that could potentially cause epidemics do so due to short-term temperature changes (Hovmoller *et al.*, 2008; Milus *et al.*, 2009). For example, soybean rust thrives in temperatures ranging from 12°C to 25°C, but exposure to just one hour at 37°C can halt its development (Bonde *et al.*, 2012).

The expected rise in global temperatures is likely to shift the regional distribution where crops are vulnerable to specific pathogens. Outside the tropics, there's a growing trend of increased pathogen survival through the winter, leading to potentially more frequent and severe outbreaks in the upcoming growing seasons. This is especially true for pathogens with special survival structures that can withstand cold, heat, or dry conditions with some surviving for several years in harsh environments.

The impact of temperature on disease development post-infection varies based on the specific host-pathogen interaction. Disease progression is typically fastest when the temperature is optimal for the pathogen but suboptimal for the host. Temperature primarily affects disease development by influencing various stages of pathogenesis, including spore germination, host penetration, pathogen growth, invasion, and sporulation. Extreme temperatures, either too high or too low for the pathogen or near optimal for the host, slow down disease development. Most fungal, bacterial, and nematode pathogens have similar durations for completing an infection cycle. The length of an infection cycle determines the number of cycles per crop season. With each cycle, the inoculum multiplies exponentially, leading to severe epidemics. Overall, favourable moisture and temperature conditions are essential and often act synergistically in initiating and developing the majority of plant diseases.

3. Effect of wind

Wind primarily impacts plant diseases by facilitating the spread of wind-borne pathogens, causing wounds on host plant surfaces, and hastening the drying of moist plant surfaces. Pathogens that spread quickly via wind or insects carried by wind, such as viruses, are more likely to cause epidemics. While basidiospores, conidia, and zoosporangia of fungi are fragile and typically do not survive long-distance wind transport, urediospores of rusts and conidia of many fungi can travel long distances across oceans by wind and remain viable. When accompanied by rain, wind-blown rain becomes more hazardous, aiding in the release and airborne transport of spores and bacteria from infected surfaces, ultimately depositing them onto wet plant surfaces. Additionally, the combination of wind and sunlight influences the speed at which plant surfaces dry.

4. Effect of light

The intensity and duration of light can either enhance or reduce the susceptibility of hosts and affect disease severity in various cases. Reduced light intensity commonly leads to the production of etiolated plants, increasing their vulnerability to non-obligate pathogens like *Fusarium* in tomatoes. However, it reduces susceptibility to obligate pathogens such as wheat stem rust. Diminished light intensity makes plants more prone to viral infections, but after infection, it tends to obscure symptom development.

5. Effect of CO_2 concentration

Current atmospheric CO_2 concentration has surpassed the 400 ppm threshold (from less than 285 ppm at the start of the 19th century (Etheridge *et al.*, 1996). The impact of CO_2 concentration on plant disease can be either positive or negative. Increased CO_2 levels can stimulate plant biomass production, leading to higher carbohydrate content that serves as food for plant-pathogenic organisms. Elevated carbohydrate levels in host tissues promote the growth of biotrophic fungi like rust, powdery mildew, and downy mildew. Higher CO_2 levels can also increase plant density, resulting in a larger leaf surface area that regulates temperature and creates a microclimate around the plant canopy. This can elevate humidity levels, thereby increasing susceptibility to foliar pathogens such as rust and powdery mildew fungi. Plants grown in high CO_2 environments can alter leaf chemistry, facilitating entry for plant pathogens through natural openings. Research by Lake and Wade (2009) on *Arabidopsis* identified more stomata in resistant varieties and fewer in susceptible ones. Elevated CO_2 levels and ozone (O_3) made resistant varieties more susceptible to powdery mildew fungus. Furthermore, increased CO_2 levels can directly impact pathogen growth. For instance, Chakraborty *et al.*, (2000) found that the growth rate of germ tubes, appressoria, and conidium ppm).

In the field conditions, plants and pathogens encounter multiple environmental factors. Understanding how these combined factors influence the outcome of disease is one of the most complex and intriguing questions in the study of plant-pathogen interactions.

Impact of climate change in plant pathology

Many sensitive regions of the world, the occurrence and severity of plant disease outbreaks are increasing, posing serious and growing dangers to primary productivity, global food security, and biodiversity loss. A wide variety of pathogens, such as bacteria, fungi, oomycetes, viruses, and nematodes, can infect plants. These pathogens differ in their target plant tissues (such as xylem, phloem, roots, or leaves), their modes of infection (from intracellular to extracellular), and their lifestyles (from biotrophs, which obtain nutrients from living cells, to necrotrophs, which obtain nutrients from dead cells). Crop plant diseases will presumably rise as a result of climate change. Initially in recent decades, globalization and international trade have increased crop pathogen mobility between continents, raising the danger of disease transmission from disease-prevalent to disease-free regions. In the new geographic area, plant species or cultivars that have not coevolved with the introduced pathogen are likely to promote pathogen prevalence and disease outbreaks. The impact of climate change in disease occurrence are enlisted:

1. Plant – pathogen interaction

The environment plays a crucial role in the growth and development of plants, with climate and weather being the most influential factors. These conditions affect various aspects such as growth stage, succulence, and genetic susceptibility in host plants, as well as the survival, vigor, and rate of multiplication of pathogens. In arid, semi-arid, moist, and sub-humid regions, temperature, rainfall, and relative humidity are the key weather factors determining disease outbreaks. Plant diseases are less severe in dry areas but are often caused by powdery mildew fungi and viruses transmitted by insect vectors. Predicting the occurrence of diseases relies on understanding relevant weather parameters and pathogen biology, aiding farmers in making timely decisions regarding control measures. Disease forecasting models are integral to integrated disease management programs. While each disease has specific requirements, plant diseases are more prevalent and severe in humid areas with cool, warm, or tropical temperatures, intermittent rains, cloudiness, and snow. Moisture, temperature, wind, light, and sunshine hours are the primary weather factors influencing disease initiation and development (Agrios, 2005).

Climate change can influence the interactions between plants and pathogens, potentially favoring the development of more virulent pathogen strains or increasing the susceptibility of host plants to infection. Changes in plant physiology, such as altered growth rates or stress responses, can also affect disease susceptibility and severity. The outcome of infection processes, the degree of disease, and the productivity of the plant are all impacted by intimate interactions between the plant, the environment, the soil and plant microbiomes, and invasive pathogens.

Pathogen evolution is driven by environmental change and human actions (e.g., global commodity and climate change), which has raised disease hazards to worldwide crops. Modern agriculture's high planting density and genetically homogenous crop monocultures have sped up the emergence of virulent diseases that can overcome disease-resistant crop types and increased the pathogen's genetic variability and population size. In a comparable way, an over dependence on pesticides has aided in the quick establishment of novel disease strains.

2. Shifts in disease distribution

Climate change alters the geographic ranges of plant pathogens and their vectors. Warmer temperatures can expand the range of pathogens into new areas previously unsuitable for their survival. Conversely, some pathogens may decline in areas where temperatures become too warm or dry for their survival. Wilt disease of banana, also known as Panama disease, caused by the soil-borne fungus *Fusarium oxysporum* f. sp. *cubense*, which likely originated in Southeast Asia and then spread globally during the twentieth century. Pathogen adaptability appears to be altered by climate change when diseases arise outside of their known geographic range or become more active within one. An example of this would be a significant *Diplodia sapinea* epidemic on *Pinus sylvestris* in Sweden.

3. Altered disease dynamics

Changes in temperature and humidity can affect the development, reproduction, and spread of plant pathogens. For example, warmer temperatures can accelerate the life cycles of pathogens, leading to increased disease incidence and severity. Changes in precipitation patterns may also affect the availability of moisture required for pathogen growth and dissemination. For instance, an extended drought stresses forest trees' water supply, making them more vulnerable to infections by pathogens that cause Phytophthora dieback disease. This could lead to the emergence of novel illnesses. The high humidity and temperature during harvest causes some mycotoxins, as fusarium mycotoxins (made by *Fusarium* spp.), to have higher quantities. Weeds proliferate more readily in humid environments, and their biomass rises as temperatures rise.

Over the past 50 years, there has been a noticeable decrease in the amount of snow cover in the Northern Hemisphere due to rising temperatures. Such alterations have the potential to cause significant community change and are important for understanding the long-term dynamics of snow blight fungus. Numerous fungal infections that propagate from plant to plant beneath the winter snowpack in many alpine and boreal plant groups are the source of snow blights.

4. Increased pathogen diversity

Climate change can lead to the introduction and establishment of new pathogens in certain regions. This can result from the migration of pathogens from other areas or the adaptation of existing pathogens to new environmental conditions. Increased pathogen diversity may pose challenges for plant disease management and control. For example, when environmental factors favor pathogen replication and susceptible hosts are available, a variety of soil opportunistic pathogens can trigger disease outbreaks. For example, soybean and wheat are extensively grown in high-density monocultures, and their yields are compromised by a plethora of pests and pathogens. Soybean rust caused by the fungus *Phakopsora pachyrhizi* and wheat blotch caused by the fungus *Zymoseptoria tritici* are among the most destructive diseases on these crops, and yield losses of more than 50% have been documented during severe epidemics. Despite the complexity of natural ecosystems (for example, biodiversity interactions), climate change and the linked emergence and evolution of pathogens pose similar challenges for wild plant communities and productivity.

5. Effect on vector borne diseases

Climate change can alter the geographic range and distribution of vector species such as insects, mites, nematodes, and other organisms that transmit plant diseases. Warmer temperatures, altered precipitation patterns, and changes in humidity can influence the habitats and survival rates of these vectors, potentially expanding their range into new territories where plants may have little to no resistance to the diseases they carry. Changes in temperature and humidity can affect the life cycle, reproduction rates, and behaviour of vector species, thereby influencing the transmission dynamics of plant diseases. For example, warmer temperatures may accelerate the development of pathogens within vectors, leading to increased rates of disease transmission. Similarly, altered precipitation patterns can affect vector breeding habitats, potentially increasing the abundance of vectors and the spread of diseases.

Climate can substantially influence the development and distribution of vectors. Changes may result in geographical distribution, increased overwintering, changes in population growth rates, increases in the number of generations and increased risk of invasion by migrant pests. Climate change can lead to more favourable conditions for the proliferation and spread of plant pathogens. Higher temperatures and humidity levels can promote the growth and reproduction of pathogens, as well as weaken plant defences, making them more susceptible to infection. This can result in higher incidences of disease outbreaks and increased severity of symptoms, leading to reduced crop yields and economic losses for farmers.

Kido *et al.* (2008) in their studies on *Melon necrotic spot virus* concluded that expression of systemic symptoms increase as temperature falls from 25 to 20°C and decrease as temperature increases from 20 to 25°C. Hence, they concluded that low temperature leads to expression of symptoms while as high temperature leads to latent infection or "Heat Masking".

6. Impact on defence mechanism in plants

Plants have developed diverse defence mechanisms to combat pathogens, including PAMP-triggered immunity (PTI), effector-triggered immunity (ETI) and sophisticated regulation of defence hormone pathways. These mechanisms are also influenced by various environmental factors.

Environmental modulation of PAMP-triggered Immunity (PTI)

Pathogen-associated molecular patterns (PAMPs) or microbe-associated molecular patterns (MAMPs) serve as distinctive molecular signatures of various microbes. These signatures play a crucial role in triggering the plant's defence mechanisms by interacting with pattern recognition receptors (PRRs) localized on the plant's plasma membrane. This recognition process is pivotal in preventing the proliferation of numerous nonpathogenic microbes that plants naturally come across. Recognition of MAMPs initiates a signaling cascade involving protein phosphorylation, the generation of reactive oxygen species (ROS), elevation in calcium ion (Ca^{2+}) levels, and the activation of genes. These events ultimately result in the inhibition of microbial growth (Agrios, 2005).

PTI signaling is altered by climatic conditions. Stomata serve as entry sites for leaf pathogens. Upon activation of PTI, stomatal closure occurs subsequent to the recognition of MAMPs by guard cells, forming a vital component of the plant's defence strategy against bacterial infiltration into the leaf apoplast. Recent findings indicate that under high humidity conditions, PTI-induced stomatal closure is impeded in both Arabidopsis and bean plants. This observation suggests that environmental factors regulate PTI-mediated stomatal defence, highlighting another layer of environmental influence on plant immunity (Panchal *et al.*, 2016).

Similarly, temperature fluctuations have the potential to modify PTI signaling. In Arabidopsis, brief exposure to elevated temperatures (28°C for 15 minutes) resulted in increased phosphorylation of PTI-associated MAPK (mitogen-activated protein kinase) and elevated expression of PTI marker genes following exposure to PAMPs. These findings imply that PTI activation may be heightened under warmer temperature conditions (Cheng *et al.*, 2013).

Environmental modulation of Effector Triggered Immunity (ETI)

One of the most potent forms of genetically mediated resistance against plant pathogens is ETI (Effector-triggered immunity). During ETI, a plant resistance (R) protein, with the majority being NLR (nucleotide-binding domain and leucine-rich repeat) proteins, detects a pathogen effector molecule or its activity within plant cells that promotes virulence. This recognition initiates a signaling cascade typically leading to a hypersensitive response (HR), a form of programmed cell death aimed at confining the pathogen at the infection site. Although effective against biotrophic pathogens, ETI proves ineffective against necrotrophs (Agrios, 2005).

High temperature causes a suppressive effect on ETI. For instance, the tobacco N protein, which provides resistance against *tobacco mosaic virus* (TMV) (Whitham *et al.*, 1996), several tomato Cf proteins offering resistance against the leaf mold pathogen *Cladosporium fulvum* (de Jong *et al.*, 2002), and several R proteins conferring resistance to the bacterial pathogen *Pseudomonas syringae* (such as RPM1, RPS2, and RPS4 from Arabidopsis), fail to initiate an efficient ETI response at temperatures exceeding 30°C (Wang *et al.*, 2009). Unlike the majority of R proteins, Xa7, a rice disease resistance protein targeting *Xanthomonas oryzae*, exhibits greater effectiveness at elevated temperatures compared to lower temperatures (Webb *et al.*, 2010).

Similar to high temperatures, elevated humidity levels can also disrupt ETI-associated hypersensitive response (HR). For instance, the reaction of tomato Cf R proteins to *C. fulvum* Avr4 and Avr9 effectors significantly diminishes when air humidity exceeds 95% (Wang *et al.*, 2005). Likewise, under high humidity, the HR associated with ETI becomes compromised for the recognition of the bacterial effector AvrRpt2 by RPS2 in *Arabidopsis* (Xin *et al.*, 2016).

Environmental influence on defence hormone pathways

Stress hormones, notably abscisic acid (ABA), salicylic acid (SA), jasmonic acid (JA), and ethylene (ET), serve as key regulators of plant responses to both abiotic and biotic factors. ABA, for instance, plays a significant role in modulating various abiotic stresses whereas SA, JA, and ET are primarily involved in orchestrating plant responses against pathogens.

Huot *et al.* (2017) reported the impact of elevated temperatures on SA-mediated defence in the interaction between *Arabidopsis* and *P. syringae.* Unlike the typical high levels of SA accumulation induced by *P. syringae* at 23°C, *Arabidopsis* failed to accumulate SA in response to *P. syringae* infection at 30°C. The biosynthesis of SA triggered by pathogens, initiates with the conversion of chorismate to isochorismate catalysed by *ICS1* (Isochorismate Synthase 1), which is then used to synthesize SA. The absence of SA accumulation was associated with the lack of expression of a significant subset of SA-responsive genes, including the *ICS1* gene. In contrast to the detrimental impact of elevated temperatures on SA defence, subjecting *Arabidopsis* plants to prolonged cold temperatures (4°C) for over a week resulted in heightened SA levels, upregulation of SA-regulated defence genes, and enhanced resistance to *Pseudomonas* infection (Kim *et al.*, 2017).

7. Impact on disease management practices

Host management

Host resistance may change due to changes in host morphology, physiology, nutrients and water. Cultivar resistance to pathogens may become more effective because of increased static and dynamic defences from changes in physiology, nutritional status, and water availability.

Durability of resistance may be threatened, however, if the number of infection cycles within a growing season increases because of one or more of the following factors: increased fecundity, more pathogen generations per season, or a more suitable microclimate for disease development. This may lead to more rapid evolution of aggressive pathogen races.

Browder and Eversmayer reported that wheat *Puccinia recondita* host pathogen gene pair related to resistance to different temperature ranges pairing produces low infectious rate. High CO_2 concentration in wheat had an average 14% reduction in nitrogen concentration in its shoot tissue that was associated with decrease susceptibility to powdery mildew. In contrast, lignification of cell walls increased in forage species at higher temperature to enhance resistance to fungal pathogen.

Chemical control

Climate change affects the efficacy of crop protection. Fungicide and bactericide efficacy can be changed by increase of CO_2, moisture, temperature, more rainfall. Changes in temperature and precipitation may alter the dynamics of fungicide residues on the crop foliage. Globally, climate change with an increased frequency of intense rainfall events, which could result in increased fungicide wash-off and reduced control.

Increase canopy coverage negatively affect spray coverage. Increase in thickness of epicuticular wax layer on leaves could result in slower and reduced uptake by host plant. Hunsche (2006) studied the influence of leaf surface characteristics on retention of mancozeb on apple and bean seedlings. He found that fungicide retention of fungicide has strong negative correlation with surface roughness and total cuticle wax. He concluded that increased canopy size could have a negative impact on spray coverage and dilution of active ingredients.

Challenges for disease management

Climate change presents challenges for the management and control of plant diseases. Climate change can disrupt the intricate balance between plants and pathogens by influencing their physiological responses and interactions. For example, elevated levels of atmospheric carbon dioxide (CO_2) can directly affect plant physiology, potentially altering their susceptibility to diseases. Additionally, changes in temperature and precipitation can modify the timing of plant phenology (e.g., flowering, fruiting), which may impact the timing of pathogen infections and disease outbreaks.

Traditional disease management strategies may become less effective as environmental conditions shift. Integrated approaches that consider climate factors, host resistance, cultural practices, and use of chemical and biological controls may be needed to mitigate the impact of climate change on plant health.

Emerging concerns on plant diseases due to climate change

The shifting climate patterns pose a significant threat to global food security by potentially worsening existing vulnerabilities in several ways. This includes the aggravation of prevalent plant diseases and the creation of favourable weather conditions for the emergence of devastating new diseases in crucial food-producing areas. Considering the current trajectory of climate change and how plants and pathogens respond to key climatic factors, certain diseases are anticipated to become significant concerns for global food security in the foreseeable future.

Climate change may facilitate the emergence of novel plant diseases or the re-emergence of previously controlled diseases. This can occur through various mechanisms, including changes in host range, altered vector biology, or the introduction of exotic pathogens into new environments. The quality and productivity of food crops, as well as the introduction and reemergence of novel pests and illnesses, are all strongly impacted by climate change. Pests and novel diseases have a detrimental effect on population social and political elements.

The Ug99 race of stem rust, which is brought on by *Puccinia graminis f. sp. tritici*, is a danger to stem rust resistance owing to Sr31, which is also a result of climate change. Higher temperatures and CO2 concentrations are also perceived as a greater threat by rice diseases including blast *(Pyricularia oryzae)* and sheath blight *(Rhizoctonia solani)*, as well as late blight *(Phytophthora infestans),* which affects potatoes.

Increased occurrence of plant diseases: In Asia, the escalation of rice bacterial and sheath blight is anticipated with rising temperatures. Similarly in Europe, sugar beet production will get affected due to heightened occurrences of leaf spot and rhizomania. In Brazil, elevated temperatures may amplify sugarcane susceptibility to *Colletotrichum falcatum* infection (Velasquez *et al.*, 2018).

Pathogen overwintering: In the United States, warmer temperatures could facilitate the overwintering of Asian soybean rust in northern latitudes, alongside promoting the development of grey leaf spot in maize. In China, warm winters have already contributed to severe outbreaks of wheat head blight in recent times (Velasquez *et al.*, 2018).

Emergence of novel pathogenic strains: The emergence of temperature-adapted races could lead to altered geographical distributions of late blight and stripe rust diseases.

Proliferation of aggressive disease vectors: Elevated populations of whiteflies transmitting viral diseases to cassava and yam are expected in Western Africa, potentially exacerbating disease spread (Velasquez *et al.*, 2018).

Increased occurrence of moisture dependent plant diseases: In Europe, this trend may pose a significant risk to potato cultivation due to the increased prevalence of late blight.

Reduced crop quality and marketability: There is a potential for a broader occurrence of mycotoxin-producing fungi in essential food crops, affecting their market value and safety (Velasquez *et al.*, 2018).

Climate change and India

India is the most affected country due to climate change. India's diverse agroecological zones and cropping systems make it particularly vulnerable to the effects of climate change on plant diseases. Changes in temperature and precipitation can affect the suitability of different crops and cropping systems, altering the diversity and distribution of pathogens and their hosts. The impact of climate change on plant diseases in India has significant implications for food security, as agriculture remains a primary source of livelihood for millions of people. Crop losses due to diseases can reduce yields, threaten livelihoods, and exacerbate food insecurity, particularly among smallholder farmers and vulnerable communities.

There is a paradigm shift in nature, time and type of occurrence of viral and other diseases of various horticultural crops due to climate change. There is a severe occurrence of *Indian Cassava Mosaic Virus* in Kerala due to shift in climate conditions and new report of *African Cassava Mosaic Virus* and *Sri Lankan Cassava Mosaic Virus* because of rise in temperature and carbon dioxide levels.

Conclusion

Climate change can have positive or negative effect on crops and their diseases. Overall, climate change is expected to have profound effects on plant pathology, with implications for agriculture, forestry, natural ecosystems, and global food security. Adaptation strategies that address the complex interactions between climate, pathogens, and host plants will be essential for sustaining plant health in a changing environment. There has been limited research on impact of climate change on plant diseases. Modified chemical and biological control needed to be implemented against diseases in the changing climate scenario. It is a responsibility of scientists to develop resistance towards various biotic and abiotic stresses.

References

Agrios, G. N. (2005). Plant pathology. Elsevier

Bonde, M. R., Nester, S. E., & Berner, D. K. (2012). Effects of daily temperature highs on development of Phakopsora pachyrhizi on soybean. Phytopathology, 102(8), 761-768.

Chakraborty, S., Tiedemann, A. V., & Teng, P. S. (2000). Climate change: potential impact on plant diseases. Environmental pollution, 108(3), 317-326.

Cheng, C., Gao, X., Feng, B., Sheen, J., Shan, L., & He, P. (2013). Plant immune response to pathogens differs with changing temperatures. Nature communications, 4(1), 2530.

de Jong, C. F., Takken, F. L., Cai, X., de Wit, P. J., & Joosten, M. H. (2002). Attenuation of Cf-mediated defence responses at elevated temperatures correlates with a decrease in elicitor-binding sites. Molecular plant-microbe interactions, 15(10), 1040-1049.

Etheridge, D. M., Steele, L. P., Langenfelds, R. L., Francey, R. J., Barnola, J. M., & Morgan, V. I. (1996). Natural and anthropogenic changes in atmospheric CO2 over the last 1000 years from air in Antarctic ice and firn. Journal of Geophysical Research: Atmospheres, 101(D2), 4115-4128.

Horino, O., Mew, T.W., and Yamada, T. (1982). The effect of temperature on the development of bacterial leaf blight on rice. Ann. Phytopath. Soc. Japan 48, 72–75.

Hovmoller, M. S., Yahyaoui, A. H., Milus, E. A., & Justesen, A. F. (2008). Rapid global spread of two aggressive strains of a wheat rust fungus. Molecular ecology, 17(17), 3818-3826.

Huot, B., Castroverde, C. D. M., Velásquez, A. C., Hubbard, E., Pulman, J. A., Yao, J., ... & He, S. Y. (2017). Dual impact of elevated temperature on plant defence and bacterial virulence in Arabidopsis. Nature communications, 8(1), 1808.

Islam, T. M., & Toyota, K. (2004). Effect of moisture conditions and pre-incubation at low temperature on bacterial wilt of tomato caused by Ralstonia solanacearum. Microbes and environments, 19(3), 244-247.

Johansen, T. J., Dees, M. W., & Hermansen, A. (2015). High soil moisture reduces common scab caused by Streptomyces turgidiscabies and Streptomyces europaeiscabiei in potato. Acta Agriculturae Scandinavica, Section B—Soil & Plant Science, 65(3), 193-198.

Jones, L. M., Koehler, A. K., Trnka, M., Balek, J., Challinor, A. J., Atkinson, H. J., & Urwin, P. E. (2017). Climate change is predicted to alter the current pest status of Globodera pallida and G. rostochiensis in the United Kingdom. Global Change Biology, 23(11), 4497-4507.

Kim, Y. S., An, C., Park, S., Gilmour, S. J., Wang, L., Renna, L., ... & Thomashow, M. F. (2017). CAMTA-mediated regulation of salicylic acid immunity pathway genes in Arabidopsis exposed to low temperature and pathogen infection. The Plant Cell, 29(10), 2465-2477.

Lake, J. A., & Wade, R. N. (2009). Plant–pathogen interactions and elevated CO2: morphological changes in favour of pathogens. Journal of experimental botany, 60(11), 3123-3131.

Magarey, R. D., Sutton, T. B., & Thayer, C. L. (2005). A simple generic infection model for foliar fungal plant pathogens. Phytopathology, 95(1), 92-100.

Mangrauthia, S. K., Singh Shakya, V. P., Jain, R. K., & Praveen, S. (2009). Ambient temperature perception in papaya for papaya ringspot virus interaction. Virus genes, 38, 429-434.

Mariette, N., Androdias, A., Mabon, R., Corbiere, R., Marquer, B., Montarry, J., & Andrivon, D. (2016). Local adaptation to temperature in populations and clonal lineages of the Irish potato famine pathogen Phytophthora infestans. Ecology and Evolution, 6(17), 6320-6331.

Milus, E. A., Kristensen, K., & Hovmøller, M. S. (2009). Evidence for increased aggressiveness in a recent widespread strain of Puccinia striiformis f. sp. tritici causing stripe rust of wheat. Phytopathology, 99(1), 89-94.

Panchal, S., Chitrakar, R., Thompson, B. K., Obulareddy, N., Roy, D., Hambright, W. S., & Melotto, M. (2016). Regulation of stomatal defence by air relative humidity. Plant physiology, 172(3), 2021-2032.

Velasquez, A. C., Castroverde, C. D. M., & He, S. Y. (2018). Plant–pathogen warfare under changing climate conditions. Current biology, 28(10), R619-R634.

Wang, C., Cai, X., & Zheng, Z. (2005). High humidity represses Cf-4/Avr4-and Cf-9/Avr9-dependent hypersensitive cell death and defence gene expression. Planta, 222, 947-956.

Wang, Y., Bao, Z., Zhu, Y., & Hua, J. (2009). Analysis of temperature modulation of plant defence against biotrophic microbes. Molecular plant-microbe interactions, 22(5), 498-506.

Webb, K. M., Ona, I., Bai, J., Garrett, K. A., Mew, T., Vera Cruz, C. M., & Leach, J. E. (2010). A benefit of high temperature: increased effectiveness of a rice bacterial blight disease resistance gene. New Phytologist, 185(2), 568-576.

Whitham, S., McCormick, S., & Baker, B. (1996). The N gene of tobacco confers resistance to tobacco mosaic virus in transgenic tomato. Proceedings of the National Academy of Sciences, 93(16), 8776-8781.

Xin, X. F., Nomura, K., Aung, K., Velásquez, A. C., Yao, J., Boutrot, F., ... & He, S. Y. (2016). Bacteria establish an aqueous living space in plants crucial for virulence. Nature, 539(7630), 524-529.

7

Mitigation and Adaptation Strategies for Climate Change and Breeding of Plants

Arya S. Nair[1] and Pramod Ashok Pimpale[2]

[1]Department of Genetics and Plant Breeding, College of Agriculture Vellayani, Thiruvananthapuram, Kerala
[2]Department of Molecular Biology and Plant Biotechnology College of Agriculture, Vellayani, Thiruvananthapuram, Kerala

Abstract

Climate change is a big game changer for agriculture in a negative way. Land, cultivation, production, etc are severely affected by climate change. Variations in rainfall, temperature, etc leads to affect the growth of crops as well as weeds, pathogens and pests. This horrifying scenario of climate change has been a discussion topic for many years now. To manage the impacts of climate change, various aspects have to be implemented. Here, we are focussing on the impact of Plant breeding on climate change. Plant breeding is basically the breeding or designing of crops according to the needs of human kind. Different methods are there for developing new varieties. So now, world is demanding for super varieties which can withstand the harsh conditions and yield better without compromising food security. The methods and examples are discussed here.

Keywords: *Climate change, Crop Improvement, Plant Breeding, Plant Biotechnology*

Introduction

Rising temperature, decreasing precipitation, increased frequency in drought and flooding, etc. is making us realize *climate change* is not some cooked up story. Still there are people who believes climate change is not real. By definition, Climate change is a long-term shift in global or regional climate patterns. This will affect the life of people in many forms: health, food, water, etc. (Table 1). Agriculture is a highly vulnerable field to such weather and climate changes, they can have significant direct and indirect impacts on farm productivity and profitability. Climate change will have adverse effect on crop physiology, growth and chemistry and thus it will cause negative impact on agricultural production. Because of the variation in climate, moisture availability may change, which limits vegetation activity/plant growth, increase in drought/heat/flooding will the crop growth and yield. Elevated CO_2 does not have a profound effect on C_4 plants. On the contrary, tissue nitrogen concentrations are higher for C_3 species under increased CO_2. Intergovernmental Panel on Climate Change (IPCC) estimates that climate change and elevated CO_2 is expected to increase competitiveness of invasive weed. IPCC also estimates that with 1°C rise in global temperature, 5 to 10 per cent yield reduction of major food and cash crops (Pachauri *et al.*, 2014).

In order to mitigate the ill-effects of climate change, various strategies are being adapted. From the perspective of crop improvement, *Super varieties* (varieties which can withstand the harsh climatic conditions) can be developed for all the crops. Varieties can be developed either through conventional plant breeding methods or through genetic engineering.

Table 1: Climate change drivers and impacts

Climate change drivers	**Climate change impacts**
Rising temperature	Suitable production habitat
Sea level rise	Pests and pollinators
Extreme events	Air, soil and water quality
Increased storms and cyclones	Distribution shifts
Droughts and floods	Altered phenology
Increased climate variability	Food nutritional quality
Decreased precipitation	Crop and animal yields
Acidification increases	Plant and animal fertility
Oxygen decreases	Grassland quality
	Tree mortality
	Heat stress
	Ecosystem degradation
	Biodiversity loss

(Taken from IPCC Sixth Assessment Report, Chapter 5)

Climate change and traditional plant breeding

Conventional plant breeding is the development or improvement of cultivars using conservative tools by manipulating plant genome within the natural genetic boundaries of the species (Acquaah, 2015). Here, new cultivars are developed by selecting yield and its components in specific environments. Not all varieties will have stable performance across the environments. General steps in breeding are: fixing the objectives, creation of variability, selection, evaluation and cultivar release. Major breeding methods are selection and hybridization. Generally, for improving one or few traits, backcross method is used commonly. Classical breeding methods can improve specific traits, so that it is effective in developing disease and stress resistance. Similarly, hybridization can utilize the potential of increased heterosis. Through thorough screening, the genotypes with desired qualities can be identified. Either they can be released as a variety or the trait can be transfer to cultivated gene pool through hybridization. Maranna *et al.* (2021) released a soybean variety NRC-128 after meticulous screening for yield and waterlogging tolerance.

Landraces are the significant source of specific traits, especially comes to stress tolerance. Evidence are available which supports the hypothesis that landraces provide sources of yield aspects and stress tolerance (Lopes *et al.*, 2015).

Another important term in classical breeding is pre-breeding. Pre-breeding refers to transferring advantageous genes from wild or exotic plant types into accepted breeding material or agricultural backgrounds. Pre-breeding plays a major role in developing crops that are resilient to variation in climate. This process involves identifying, collecting, and utilizing genetic resources especially wild species to enhance crop quality and productivity. Pre-breeding strategies help to exploit wild or closely related crop species that exhibit desirable traits, such as heat or drought tolerance, to create novel robust varieties against environmental stress (Sukumaran *et al.*, 2021). The process of pre-breeding is also time taking. The primary goals of pre-breeding are to offer readily available genetic material for future plant breeding programmes. Even though conventional breeding methods are successful, it is very tedious and difficult. Variety development through traditional plant breeding methods will take atleast 6-8 years. Participatory breeding and participatory varietal selection may help in fastening the development of climate-resilient crop varieties and cropping systems aimed at multiple breeding targets (banga and Kang, 2014). The emergence of new breeding approaches such as Rapid Generation Advance (RGA), Doubled Haploid (DH), Shuttle Breeding, Marker Assisted Back Crossing (MABC), Marker Assisted Recurrent Selection (MARS), Genomic Selection (GS), Speed Breeding, etc. also can been used to shorten the breeding cycle along with efficient screening for specific biotic and abiotic stresses. Hence, accelerated breeding is a systematic tool for developing new varieties in a shorter period of time, to reduce the impact of climate change (Atlin *et al.*, 2017). Similarly, Genetically Modified crops, Gene editing, etc are also effective for developing climate smart varieties.

Climate change and new breeding approaches

i) Molecular breeding techniques

The development of molecular breeding techniques has revolutionized crop improvement field. Those techniques have empowered breeders to precisely select desired traits at the molecular level. The vast knowledge of plant genomes and the identification of genes associated with stress tolerance is a pre-requisite for utilizing these techniques. Use of molecular markers is the headway in this mechanism. These markers are the specific regions in the genome that can be easily detected and linked to particular traits of interest. The concept of associating molecular marker to a quantitative trait is suggested by Sax (1923). Molecular markers can be used in plant breeding programmes for germplasm characterization, genetic diversity assessment and identification of crop varieties via DNA fingerprinting (Farooq and Azam, 2002). Marker-assisted selection (MAS) allows breeders to identify and select plants with desired stress tolerance traits more efficiently. If the genetic markers associated with stress tolerance are identified, breeders can accurately predict the performance of potential offspring and select the most promising ones for further breeding. The desirable traits from wild crop relatives and landraces for stress tolerance can be introgressed into cultivated crop varieties through conventional hybridization techniques. Flawless phenotyping and genotyping will ensure the transfer specific stress-tolerance genes from these genetic resources into elite cultivars. This broadens the available genetic diversity for breeding programs, thereby enhancing the potential for developing climate-resilient crops (Hafeez *et al.*, 2023).

For example, by using marker-assisted breeding approaches, several QTLs for drought tolerance in rice is incorporated into elite cultivars. Singh *et al.*, 2016 have successfully incorporated QTLs such as *qDTY9.1*, *qDTY2.2*, *qDTY10.1* and *qDTY4.1* in high-yielding variety IR64 by marker-assisted backcrossing approach. Drought-tolerant elite Malaysian rice cultivar MR219 was developed with the pyramiding of three QTLs, *qDTY2.2*, *qDTY3.1* and *qDTY12.1* (Shamsudin *et al.*, 2016) . Similarly rice variety TDK1 was developed for high yield under drought through the incorporation of three QTLs (*qDTY3.1*, *qDTY6.1* and *qDTY6.2*) (Dixit *et al.*, 2017).

QTLs of various traits associated to the stress tolerance were reported for different stress and different crops. If we can pinpoint QTLs, those can be introgressed into desired elite varieties. As drought tolerance, submergence tolerance and various disease tolerances were incorporated into different varieties of different crops utilizing molecular marker techniques. For marker assisted techniques, QTLs and closely linked markers are important. For mapping qtls, high-throughput phenotyping is also required.

ii) GWAS and genomic selection

The advances in the field of sequencing and phenotyping helped GWAS to overcome the constraints of QTL mapping (Huang and Han, 2014). Genome-wide association studies (GWAS) helps to identify associations of genotypes with phenotypes by testing the differences in allele frequency of genetic variants between individuals who are ancestrally similar but differ phenotypically (Uffelmann *et al.*, 2021). GWAS helps in the better understanding of the genetic architecture of traits. In plants, GWAS has widespread applications reported regarding to biotic and abiotic stresses. GWAS have been applied to on drought tolerance (Thoen *et al.*, 2017), salt tolerance (Wan *et al.*, 2017), and heat tolerance (Lafarge *et al.*, 2017).

Abou-Elwafa and Shehzad (2021) performed GWAS with 15,737 SNP markers and six phenotypic traits in 290 lines of wheat. From the study, they identified 205 significant marker-trait associations (MTA) for the six phenotypic traits under well-irrigated and drought stress conditions. The knowledge on specific traits and QTLs will help in enhancing the tolerance ability and exploit in the breeding schemes.

Similarly, Warraich *et al.* (2020) evaluated 180 diverse rice accessions for salinity tolerance at reproductive stage. They genotyped them using simple sequence repeats markers covering all 12 chromosomes and identified 28 significant MTA and among them 19 associations were identified for Na^+, K^+, Na^+/K^+ uptake in stem and leaves. Mwando *et al.* (2020) detected 19 loci containing 52 significant salt-tolerance-associated markers across all chromosomes, and 4 genes belonging to 4 family functions underlying the predicted marker trait associations.

Chopra *et al.* (2017) reported different heat and cold stress tolerant genes in *Sorghum bicolor*. Through the experiment, 30 SNPs were identified for genes related to anthocyanin expression and carbohydrate metabolism. Those are associated with cold stress at the seedling growth phase of sorghum. Likewise, 12 SNPs were discovered for heat stress and controlled by the genes related to ion transport mechanism and sugar metabolism.

In soybean, GWAS was performed in relation with seedling flooding tolerance. Two major SNPs, Gm_08_11971416, and Gm_08_46239716 were found to be consistently connected with seed-flooding tolerance related traits (Sharmin *et al.*, 2021). These results provide discernment about new genetic resources and information to improve abiotic stress tolerance in future crop breeding programmes *via* genomic and marker-assisted selection and to explore functional characterization of the identified candidate genes, etc.

Genomic selection (GS) helps in the prediction of a genomic estimate of breeding value (GEBV) using genome-wide markers, which can be used to select favorable individuals. GS involves three essential steps: prediction model training, prediction of breeding value, and selection of favorable individual based on the predicted GEBV (Gidamo, 2022). Annicchiarico *et al.*, 2017 conducted a study which aims at assessment of the predictive ability of genomic selection (GS) for grain yield under severe drought using genotyping-by-sequencing (GBS) data. The results suggests that the intrapopulation predictive ability of the best-performing GS models was very high (around 0.7) for onset of flowering and yield.

The breeding values of 240 subtropical maize lines were tested by phenotyping for drought at varied environments using 29,619 cured SNPs (Shikha *et al.*, 2017). They compared the prediction accuracies of seven genomic selection models (ridge regression, LASSO, elastic net, random forest, reproducing kernel Hilbert space, Bayes A and Bayes B) and those were tested for their agronomic traits. Bayes B model outperformed the other models. From Bayes B, a set of 1053 significant SNPs was selected to validate the genes and QTLs, which were having higher marker effects. Among these 1053 SNPs, 77 SNPs showed association with 10 drought-responsive transcription factors. These transcription factors were related to different physiological and molecular functions of drought tolerance. These experiments are very much relevant for the identification and selection of superior genotypes and candidate genes for developing drought-tolerant maize hybrids (Shikha *et al.*, 2017).

iii) Speed breeding (SB)

Normally, the development of elite drought and salinity tolerant varieties through conventional approaches takes around 10–12 years due to long reproductive cycles of various crops and long selection cycles. Speed Breeding (SB) technology came as a solution that problem. Speed breeding is a collective technique which aims at reducing the reproductive period of crop plants by controlling environmental conditions. Temperature regulation, manipulation of the photoperiod regime, increasing CO_2 levels, regulating soil moisture, inducing plant growth and hormones, etc. enhance the growth in plants as well as the pace of shift between the vegetative and the reproductive phase which further helps crop plants to tolerate stress conditions. Moreover, rapid and multiple trait selection is possible through speed breeding combined with the approaches of marker-assisted selection (MAS) and high-throughput phenotyping (Rai *et al.*, 2023). SB also allows for the integration of advanced protocols such as gene editing and genotyping which accelerates crop improvement. SB Protocols for various crops, including cereals, pulses and canola crops have been developed, producing 4–6 generations in a year. With its application to a wide range of crops and lower labour requirements than breeding methods, SB offers an efficient way for crops with large populations. Speed breeding is a potent method for rapidly developing new plant varieties (Chaudhary and Sandhu, 2024).

iv) Climate change and genetically modified crops

Biotechnology research aims to change the genetic makeup of crops to enhance their resistance to environmental and biological stresses. Some examples of biotechnology in crop improvement include genetic engineering, genome editing, RNA-mediated gene silencing, and genome mapping. Biotechnology can help address climate change and its effects through improved crop quality, quantity, nutrition, taste, and shelf life. For example, biotech crops can help reduce CO_2 emissions by allowing farmers to use less fertilizer and energy, and practice soil carbon sequestration. Herbicide tolerant crops like soybean and canola can also help reduce fuel use and soil erosion.

The traits which are important for adoption in climate change are drought tolerance, salinity tolerance, heat stress tolerance, efficient water usage, pest resistance, disease resistance. Genetic modification techniques can alleviate the negative effect on climate change by degreasing greenhouse gaseous, biofuel use, carbon sequestration, less use of fertilizer and tolerance of biotic and abiotic stress. GM crops may reduce agricultural greenhouse gas emissions. GM yield increases can reduce production emissions while simultaneously mitigating land-use change and associated emissions. Globally, the most extensively accepted genetically modified traits are insect resistance (IR) and herbicide tolerance (HT) (Kovak *et. al.* 2022). These features prevent crop damage from insects and weeds, resulting in higher yields. Increased crop yields can minimise the requirement for new land for food production, reducing CO_2 emissions from land-use change.

GM herbicide-tolerant crops have also been produced to allow for the use of less toxic and ecologically friendly pesticides. Glyphosate and glufosinate (another GM targeted herbicide resistance) are both Class III herbicides (EPA), meaning they are not extremely harmful and have a short persistence in the soil and environment, averaging around 40 days. These herbicides have replaced herbicides that are more hazardous or have been shown to contaminate and stay in groundwater. Glyphosate replaced numerous Class II herbicides (very hazardous) by more than 83% when used on herbicide-tolerant soybeans. (Oliver, 2014) Thus, targeting less hazardous herbicides reduces human exposure while improving environmental and human health.

The main transgene utilised in the creation of insect resistant GM crops is one that permits the synthesis of a CRY protein toxin from the bacteria *Bacillus thurengensis (Bt).* This toxin is very selective to critical agronomic caterpillar and beetle pests that feed on crop plants, damaging the insect's stomach cells and inhibiting digestion. The CRY poisons are unique to their target insects, harmless to vertebrates (including humans), and have no effect on the plant. The introduction of Bt-GM crops has resulted in a considerable drop in the usage of chemical pesticides in all countries where they have been implemented, as well as a reduction in environmental effect and associated human exposure.

v) Gene editing for climate change in agriculture

Gene editing is a technique for creating DNA alterations at precise genomic regions. These alterations can cause the deletion or knockdown of one or more genes without the permanent inclusion of any foreign DNA. Alternatively, genes from the organism's genepool or from other animals can be introduced into specific sites in the genome to introduce a new feature. Transcription activator-like effector nucleases (TALENs), Zinc Finger Nucleases (ZFNs), and CRISPR/Cas systems have all been used to achieve precise gene editing. (Khalil, 2020). In banana, gene editing with CRISPR/Cas9 to induce knockouts of genes involved in gibberellin production has aided the development of a semi-dwarfed cultivar. (Shao *et al.*, 2020). This cultivar may be more resistant to lodging due to strong winds, typhoons, and storms, which are expected to become more severe as a result of climate change. Gene editing for a variety of rice diseases has proven incredibly effective. CRISPR/Cas9 was used to generate knockouts of OsSWEET13. Pathogens can exploit SWEET family genes, which encode sucrose transporters (Jiang *et al.*, 2013). Mutating this gene significantly enhanced disease resistance (Zhou *et al.*, 2015).

Conclusion

The increasing world population is increasing the demand for crop production. By 2050, global agricultural production may need to be doubled to meet increasing demands. Climate change will decelerate the production process. Various approaches can be utilized to manage the impacts of climate change on agriculture like development of new crop production packages, new plant protection mechanisms, etc. From plant breeding point of view, development of climate smart varieties is important. The varieties which can tolerate harsh climatic conditions and still performing better. To ensure food security and sustainability, the integrated approach of different disciplines is needed. Approaches should accelerate the crop production process, while mitigating climate change.

References

Abou-Elwafa, S.F., Shehzad, T. Genetic diversity, GWAS and prediction for drought and terminal heat stress tolerance in bread wheat (Triticum aestivum L.). Genet Resour Crop Evol 68, 711–728 (2021). https://doi.org/10.1007/s10722-020-01018-y

Annicchiarico, P., Nazzicari, N., Pecetti, L., Romani, M., Ferrari, B., Wei, Y., & Brummer, E. C. (2017). GBS-based genomic selection for pea grain yield under severe terminal drought. The Plant Genome, 10(2), plantgenome2016-07.

Banga, S. S., & Kang, M. S. (2014). Developing climate-resilient crops. Journal of Crop Improvement, 28(1), 57-87.

Chaudhary, N., Sandhu, R. A comprehensive review on speed breeding methods and applications. Euphytica 220, 42 (2024). https://doi.org/10.1007/s10681-024-03300-x

Chopra, R., Burow, G., Burke, J.J., Gladman, N., and Xin, Z. 2017. Genome-wide association analysis of seedling traits in diverse Sorghum germplasm under thermal stress. BMC Plant Biol. 17, 12.

Farooq, S., & Azam, F. (2002). Molecular markers in plant breeding-II. Some pre-requisites for use. Pakistan journal of biological Sciences, 5(10), 1141-1147.

Gidamo, G. H. (2022). Genomic Selection: A Faster Strategy for Plant Breeding. In Case Studies of Breeding Strategies in Major Plant Species. IntechOpen.

Hafeez, U., Ali, M., Hassan, S. M., Akram, M. A., & Zafar, A. (2023). Advances in breeding and engineering climate-resilient crops: A comprehensive review. International Journal of Research and Advances in Agricultural Sciences, 2(2), 85-99.

Huang X H, Han B. 2014. Natural variations and genome-wide association studies in crop plants. Annu Rev Plant Biol, 65: 531–551

Jiang, W., Zhou, H., Bi, H., Fromm, M., Yang, B., and Weeks, D. P. (2013). Demonstration of CRISPR/Cas9/SgRNA-mediated targeted gene modification in arabidopsis, tobacco, sorghum and rice. Nucleic Acids Res. 41:e188. doi: 10.1093/nar/gkt780

Khalil, A. M. (2020). The genome editing revolution: review. J. Genet. Eng. Biotechnol. 18:68. doi: 10.1186/s43141-020-00078-y

Kovak E, Blaustein-Rejto D, Qaim M. Genetically modified crops support climate change mitigation. Trends Plant Sci. 2022 Jul;27(7):627-629. doi: 10.1016/j.tplants.2022.01.004. Epub 2022 Feb 8. PMID: 35148945.

Lafarge, T.; Bueno, C.; Frouin, J.; Jacquin, L.; Courtois, B.; Ahmadi, N. Genome-wide association analysis for heat tolerance at flowering detected a large set of genes involved in adaptation to thermal and other stresses. PLoS ONE 2017, 12, e0171254

Maranna, S., Nataraj, V., Kumawat, G., Chandra, S., Rajesh, V., Ramteke, R., ... & Khandekar, N. (2021). Breeding for higher yield, early maturity, wider adaptability and waterlogging tolerance in soybean (Glycine max L.): A case study. Scientific reports, 11(1), 22853.

Mwando, E., Han, Y., Angessa, T. T., Zhou, G., Hill, C. B., & Li, C. (2020). Genome-wide association study of salinity tolerance during germination in barley (Hordeum vulgare L.). Frontiers in plant science, 11, 478446.

N.A.A. Shamsudin, B.P.M. Swamy, W. Ratnam, M.T.S. Cruz, A. Raman, A. Kumar. Marker assisted pyramiding of drought yield QTLs into a popular Malaysian rice cultivar, MR219. BMC Genet, 17 (1) (2016), p. 30

Oliver, M. J. (2014). Why We Need GMO Crops in Agriculture. Missouri Medicine, 111(6), 492-507. https://www.ncbi.nlm.nih.gov/pmc/articles/PMC6173531/

Pachauri, S. (2014). Household electricity access a trivial contributor to CO2 emissions growth in India. Nature Climate Change, 4(12), 1073-1076.

Prabhu, K. R., Kumar, A., Yumkhaibam, R. S., Janeja, H. S., Krishna, B., & Talekar, N. (2023). A review on conventional and modern breeding approaches for developing climate resilient crop varieties. Journal of Applied and Natural Science, 15(3), 987-997.

R. Singh, Y. Singh, S. Xalaxo, S. Verulkar, N. Yadav, S. Singh, N. Singh, K.S.N. Prasad, K. Kondayya, P.V. RamanaRao, M.G. Rani, T. Anuradha, Y. Suraynarayana, P.C. Sharma, S.L. Krishnamurthy, S.K. Sharma, J.L. Dwivedi, A.K. Singh, P.K. Singh, Nilanjay, N.K. Singh, R. Kumar, S.K. Chetia, T. Ahmad, M. Rai, P. Perraju, A. Pande, D.N. Singh, N.P. Mandal, J.N. Reddy, O.N. Singh, J.L. Katara, B. Marandi, P. Swain, R.K. Sarkar, D.P. Singh, T. Mohapatra, G. Padmawathi, T. Ram, R.M. Kathiresan, K. Paramsivam, S. Nadarajan, S. Thirumeni, M. Nagarajan, A.K. Singh, P. Vikram, A. Kumar, E. Septiningshih, U.S. Singh, A.M. Ismail, D. Mackill, N.K. Singh. 2016. From QTL to variety-harnessing the benefits of QTLs for drought, flood and salt tolerance in mega rice varieties of India through a multi-institutional network. Plant Sci, 242 (2016), pp. 278-287

Rai, N. K., Ravika, Yadav, R., Jattan, M., Karuna, Rai, P. S., .& Yadav, S. (2023). Speed Breeding: A Budding Technique to Improve Crop Plants for Drought and Salinity Tolerance. In Salinity and Drought Tolerance in Plants: Physiological Perspectives (pp. 295-313). Singapore: Springer Nature Singapore.

S. Dixit, A. Singh, N. Sandhu, A. Bhandari, P. Vikram, A. Kumar. Combining drought and submergence tolerance in rice: Marker-assisted breeding and QTL combination effects. Mol Breeding, 37 (2017), p. 143

Sharmin, R. A., Karikari, B., Chang, F., Al Amin, G. M., Bhuiyan, M. R., Hina, A., & Zhao, T. (2021). Genome-wide association study uncovers major genetic loci associated with seed flooding tolerance in soybean. BMC Plant Biology, 21, 1-17.

Shikha, M., Kanika, A., Rao, A. R., Mallikarjuna, M. G., & Nepolean, T. (2017). Genomic selection for drought tolerance using genome-wide SNPs in maize. Frontiers in plant science, 8, 261427.

Thoen, M.P., Davila Olivas, N.H., Kloth, K.J., Coolen, S., Huang, P.P., Aarts, M.G., Bac-Molenaar, J.A., Bakker, J., Bouwmeester, H.J., Broekgaarden, C. 2017. Genetic architecture of plant stress resistance: Multi-trait genome-wide association mapping. New Phytol. 213, 1346–1362

Uffelmann, E., Huang, Q.Q., Munung, N.S. et al. Genome-wide association studies. Nat Rev Methods Primers 1, 59 (2021). https://doi.org/10.1038/s43586-021-00056-9

Wan, H., Chen, L., Guo, J., Li, Q., Wen, J., Yi, B., Ma, C., Tu, J., Fu, T., Shen, J. 2017. Genome-wide association study reveals the genetic architecture underlying salt tolerance-related traits in rapeseed (Brassica napus L.). Front. Plant Sci. 8, 593.

Warraich, A.S., Krishnamurthy, S.L., Sooch, B.S. et al. Rice GWAS reveals key genomic regions essential for salinity tolerance at reproductive stage. Acta Physiol Plant 42, 134 (2020). https://doi.org/10.1007/s11738-020-03123-y

Zhou, J., Peng, Z., Long, J., Sosso, D., Liu, B., Eom, J.-S., et al. (2015). Gene targeting by the TAL effector PthXo2 reveals cryptic resistance gene for bacterial blight of rice. Plant J. 82, 632–643. doi: 10.1111/tpj.12838

8

Mitigation and Adaptation Strategies for Climate Change and Biotechnology

Swarup Premanand Nagrale and Dinesh Arya

Agri-Biotechnology Division, National Agri-Food Biotechology Institute Mohali, Punjab

Abstract

In the face of rising climate change and food insecurity problems, investigating the potential of microorganisms provides a viable and long-term answer. The discussion below investigates the intricate relationship between microbial groupss and the dual challenges of environmental disaster and food security. Microorganisms, ranging from bacteria to fungus and archaea, affect our planet's ecosystems by influencing soil health, nitrogen cycling, and plant-microbe interactions. It then emphasises the reciprocal effects of human-caused environmental change on microorganisms and their habitats. Microbial biotechnology have the potential to revolutionise agriculture and assure global food security, as evidenced by innovations such as biofertilizers and insecticides. The chapter highlights the symbiotic relationship between bacteria and sustainable food production. Microbial biotechnology can help agriculture adapt to changing climatic conditions, reducing water shortages and increasing soil moisture retention. Their potential to increase production in both traditional and precision agriculture under a variety of climatic situations is emphasised. Finally, chapter advocates for the immediate acknowledgment and exploitation of microbial power for a sustainable future. Adopting microbial biotechnology not only promotes environmental stewardship, but also lays the groundwork for a resilient and resource-efficient agricultural future.

Keywords: *Microorganism, Climate change, Biotechnology, food security, nitrogen cycling*

Introduction

Climate change is a deadliest hazard to life on Earth. Global temperature is rising due to greenhouse gas emissions from the usage of fossil fuels, this result in widespread agricultural and fisheries failure, the extinction of hundreds of thousands of species, and whole towns becoming uninhabitable (Lindwall, 2022). Rising global temperatures are severely killing humans and trees, forcing half of all species to move, worsening water-borne and respiratory ailments in people, and jeopardising millions' food and water security (Isaacs-Thomas, 2022). Addressing these concerns necessitates a worldwide collaboration to minimise greenhouse gas emissions and promote sustainable behaviours. Climate change affects agriculture production, lowers crop nutritional value, and increases the frequency of extreme weather events. These interruptions jeopardise food production, availability, and access, especially in vulnerable countries. Microbes play an important role in soil health, nitrogen cycling, and ecosystem stability. They impact a variety of soil activities, such as organic matter breakdown, nutrient mobilisation, and plant growth stimulation. However, climate change poses a substantial danger to microbial communities, possibly changing their activities and altering ecosystem dynamics (Wang *et al.*, 2022). Transitioning to a low-carbon economy requires a global effort. Mitigation strategies include enhancing energy efficiency, adopting renewable energy sources, electrifying industrial processes, implementing efficient transportation systems, and introducing carbon taxes or emissions markets.

The increasing use of agrochemicals in modern agriculture has created substantial hurdles for agricultural sustainability. While these pesticides have clearly helped to enhance agricultural yields, they have also had a number of negative impacts. Pesticides and fertilisers are used indiscriminately, which harms beneficial soil organisms, animals, human health, and crop nutritional value. Furthermore, abuse of pesticides has resulted in the evolution of resistant pests and diseases, worsening the problem. Furthermore, current agricultural techniques have resulted in a significant loss of crop variety, leaving agriculture more prone to insect outbreaks, disease infestations, and environmental pressures including water scarcity, severe temperatures, and rising salt. To properly handle these numerous difficulties, a comprehensive strategy is required. This strategy should include adopting sustainable agriculture methods, implementing soil conservation measures, and developing climate-resilient techniques. By accepting these ideas, we may work towards both food security and long-term ecosystem sustainability (Brzozowski & Mazourek, 2018; Damalas & Eleftherohorinos, 2011; Dhanaraju *et al.*, 2022).

To provide strong and constant agricultural yields, plants must be protected from both biological and environmental stressors. Introducing genetic variety into crops can help to battle illnesses, increase yield, improve tolerance to environmental variables, and boost biodiversity. To achieve long-term agricultural sustainability, crop types that are high in production and resistant to diseases, pests, and severe environmental circumstances such as drought, excessive temperatures, and soil salinity are critical. Traditional techniques have mostly focused on discovering organisms with high adaptability and performance. Environmental stressors, which include both biotic and abiotic factors such as chemicals, temperature, light, and other climatic variables, act as selection pressures, influencing species' acclimation, adaptation, and evolutionary processes (Bailey-Serres *et al.,* 2019; Cheng & Cheng, 2015; Dong & Ronald, 2019; Das *et al.*, 2023).

Microbial Biotechnology: An overview

Microbial biotechnology is the employing of biotechnology concepts and modern technology to research and utilise microorganisms and their products. Bacteria, fungus, and other microorganisms are used to perform a variety of functions that benefit human health, food, agriculture, industry, and the environment (licón-Hernández *et al.,* 2022). Microbial biotechnology tackles global challenges in health, food security, environmental sustainability, and industrial production by leveraging advances in genetic engineering, synthetic biology, protein omics, and high-throughput screening technologies (De Sousa *et al.,* 2018; Poblete-Castro *et al.*, 2020). For example, microorganisms are utilised to ferment a variety of industrial products, including antibiotics, enzymes, organic acids, biofuels, and medicines (Raveendra *et al.*, 2018; Mosunova *et al.*, 2021; Yafetto, 2020).

Microorganisms serve an important part in a variety of ecological processes, allowing ecosystems function properly. Their impact spans beyond ecological dynamics, as their potential applications in tackling climate change and food hunger have gained traction. Microorganisms serve a central role in nutrient cycling because they break down organic matter into vital nutrients that plants and other organisms may utilise. Decomposition is an essential mechanism for supporting life in ecosystems (Raza *et al.,* 2023). Certain bacteria, such as nitrogen-fixing bacteria found in leguminous plant root nodules, transform atmospheric nitrogen into a form that plants can use, hence increasing soil fertility (Chakraborty & Kundu, 2023). Furthermore, microorganisms play a key role in the biodegradation of pollutants (Fig. 1), that helps to reduce the impact of environmental toxins on ecosystems (Dave & Das, 2021; Hussein *et al.*, 2012). Numerous plants form symbiotic relationships with mycorrhizal fungi that aid in nutrient intake, whereas some bacteria form symbiotic connections with plant roots, offering benefits such as better resistance to environmental stress (Hussein *et al.*, 2012).

Climate change is broadly characterised as a major shift in average weather conditions, such as growing warmer, wetter, or drier over several decades or more. Microbial biotechnology is at the forefront of new climate change mitigation efforts, with solutions spanning from carbon sequestration and methane reduction to bioenergy production and circular economy practices. Several scientific investigations and technical breakthroughs in microbial biotechnology have shown the promise for sustainable and ecologically friendly techniques to combating climate change, as shown below.

Microbial biotechnology provides interesting options for mitigating climate change by utilising the specific characteristics of distinct environmental processes, such as increasing carbon sequestration, notably by bacteria in soil, through organic matter breakdown. This entails altering microbial populations to increase carbon storage in soils (Onyeaka & Ekwebelem, 2023). Methane is a powerful greenhouse gas, and methanotrophic bacteria play an important role in its oxidation. These bacteria are being studied for their potential to reduce methane emissions from both natural and anthropogenic sources (Nazaries *et al.,* 2021). Another crucial component is microorganisms' ability to produce biofuels, giving a sustainable alternative to fossil fuels. Advances in metabolic engineering and synthetic biology improve the efficiency of microbial biofuel generation (Lopes & da Silva, 2021).

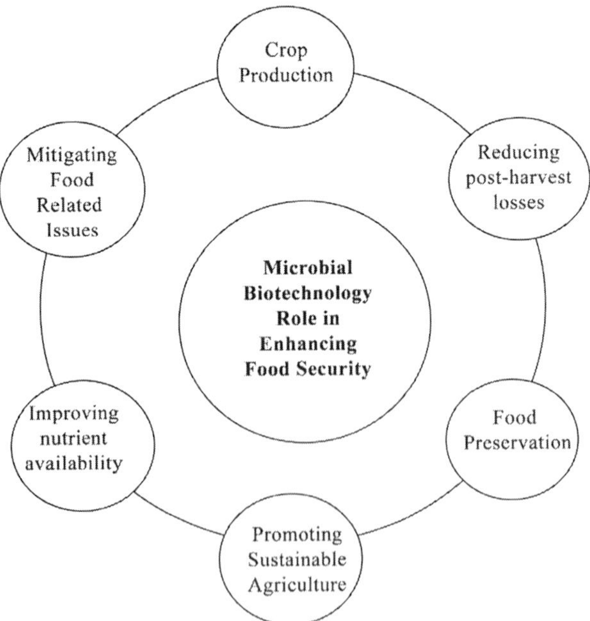

Fig. 1: Role of microbial biotechnology in enhancing food security

Microbial methods, such as anaerobic digestion, transform organic waste into biogas, which is a sustainable energy source. This helps with waste management and provides an alternative to fossil fuels (Kougia *et al.,* 2021). According to Compant *et al.* (2021), endophytic bacteria found in plant tissues are being studied for their potential to improve plant health, nutrient absorption, and resistance to diseases and environmental stresses. Microbial diversity in soil improves soil health and resilience. This, in turn, impacts carbon and nitrogen cycling, which influences greenhouse gas dynamics (Zhang *et al*., 2020).

Microbial Biotechnology for Enhancing Food Security

Microbial biotechnology contributes significantly to food security by providing new solutions to difficulties in agriculture, crop production, and food preservation, as well as encouraging sustainable agriculture, increasing crop output, and alleviating food-related concerns. Microbial biotechnology offers a diverse approach to addressing food security issues by increasing crop output, improving nutrient cycling, preserving soil health, and providing long-term pest management solutions. These microbial-based technologies provide ecologically friendly solutions, helping to establish resilient and sustainable agricultural systems. Bhattacharyya *et al.* (2020) shown that microbial biofertilizers, such as nitrogen-fixing bacteria and mycorrhizal fungi, improve nutrient availability for plants, hence boosting sustainable and environmentally friendly farming methods.

Microorganisms, such as entomopathogenic fungi and bacteria, were investigated for their potential function in biological pest control, offering alternatives to chemical pesticides and helping to sustainable pest management (Kumar *et al.,* 2021). Beneficial microbes that boost plant development and stress tolerance were also used to improve agricultural resilience in harsh environmental circumstances, resulting in higher yields (Nadeem *et al.,* 2021). Microbial biotechnology helps to food security through microbial bio preservation strategies that employ beneficial bacteria to enhance food product shelf life by preventing the growth of spoilage and harmful pathogens (Gänzle, 2015).

Some research have shown that microbial-based techniques can enhance crop productivity, nutrient cycling, soil health, and insect management, hence providing long-term answers to global food concerns. Basha *et al.* (2016) discovered that Plant development-boosting Rhizobacteria (PGPR) improved plant development by boosting nutrient absorption, creating growth-promoting chemicals, and guarding against diseases. Smith and Read (2010) discovered that arbuscular mycorrhizal fungi (AMF) create symbiotic relationships with plant roots, boosting nutrient uptake and water absorption, hence increasing agricultural yield, particularly in nutrient-deficient soils.

Furthermore, as reported by Bhattacharyya *et al.* (2020), nitrogen-fixing bacteria and other microbial biofertilizers contributed to nutrient cycling, enriching the soil with important nutrients, hence increasing soil fertility and promoting sustainable agriculture. Other successful research studies were mentioned in the review article by (Reisoglu and Aydin, 2023) about how microalgae were cultivated for bioenergy production, where algae could be used to capture carbon dioxide and produce biofuels, providing a renewable energy source while mitigating climate change.

Microbial biodiversity and adaptability

Microbial biodiversity is inextricably linked to a number of ecological concepts, including the collective presence of all organisms and abiotic elements within a specific environment, known as an ecosystem, and the specific portion of an ecosystem where a community of organisms can thrive, known as habitat. Microbes, which include a wide variety of microorganisms such as fungus, bacteria, and viruses, are essential to life on Earth, playing key roles in maintaining ecosystem processes and services.

Some research papers highlight the importance of microbial communities in functions critical to global well-being. Microbial communities serve critical roles in nutrient cycling and food security, trash decomposition, primary production, and influencing the structure of seas and atmospheres, all of which contribute considerably to climate change dynamics (Delgado-Baquerizo *et al.*, 2018; Louca *et al.*, 2016).

The vast range of microorganisms is incredible, with an estimated 5-30 million species. Only around 2 million of them have been formally documented, with the vast majority unexplored or nameless. Over 1 billion bacteria may be found in only one gramme of soil, with fewer than 5% of them recognised. The wide fungal diversity, which includes at least 1.5 million species, highlights the richness of microbial life (Tedersoo *et al.,* 2014). This microbial diversity outnumbers that of arthropods and seed plants combined, highlighting its ecological importance (Onen *et al.,* 2020). Despite the incredible richness of microbial communities, information is biased towards bigger creatures, temperate habitats, and organisms that are useful to humans.

Surprisingly, soil contains between 55% and 98% of Earth's total biodiversity, owing mostly to the existence of hundreds of microbial species and genotypes (Anthony *et al.,* 2023). Agricultural soil alone contains around 3,000 kg (fresh weight) of microorganisms per hectare, demonstrating their widespread presence (Fierer, 2017). Microorganisms have an amazing ability to adapt to varied settings, which is inherent in their genetic and reproductive capabilities. This flexibility produces significant variety, yet the specificity of selection gives them a distinct edge.

Nature regulates microbial development through a variety of parameters, including critical resources such as carbon, nitrogen, macronutrients, micronutrients, oxygen, and other electron acceptors. Temperature, water potential, pH, oxygen availability, light, and osmotic conditions all play important roles in microbial development (Thakur *et al.*, 2022). Microorganisms have the ability to adapt and flourish in extremely hostile settings. These extreme niches include hot springs (Malkawi & Al-Omari, 2010), saline lakes, dry deserts, deep ocean depths, acidic and alkaline environments, polar frigidity, polluted places, and areas with limited energy and nutritional supplies. Microorganisms in these severe environments adapt to changing circumstances by developing cellular, biochemical, and molecular methods. They produce enzymes, compounds, and metabolites that guard against salinity, pH, temperature, pressure, solar radiation, nutrient access, oxygen levels, osmotic pressures, and gravitational fluctuations (Thakur *et al.*, 2022). Diverse microorganisms respond to environmental stimuli in unique ways, honing survival systems.

Understanding and comprehending the complexities of microbial biodiversity is critical for realising their full potential in maintaining ecosystem health and solving global concerns.

Microbial Contributions to Nutrient Cycling, Food Production and Food Security

Microorganisms are frequently underestimated in terms of importance, despite the fact that they serve a critical role in ecosystem function, laying the groundwork for agriculture and food production. Several studies have emphasised the complicated network of microbial activities that impact elemental cycles such as carbon, nitrogen, sulphur, and iron, demonstrating the critical role these microbes play in supporting ecosystems, as shown below:

Microbial Impact on Elemental Cycles

Microorganisms have a substantial impact on elemental processes such as the carbon cycle, nitrogen cycle, sulphur cycle, and iron cycle (Kuypers, 2018; Luo *et al.,* 2020). Bacteria, yeast, and fungus, including well-known species like Saccharomyces cerevisiae, all play important roles in these cycles.

Microbial contribution to agriculture

Microbial contributions to agriculture are critical for ensuring sustained and productive crop production. Plant-microorganism interactions, such as those involving bacteria, fungus, and archaea, are critical for nutrient cycling, disease control, and soil health. In addition, microbial activities help to create and stabilise soil structure. This is critical for water retention, aeration, and soil health (Six *et al.*, 2004). Chen *et al.* (2023) found that the phyto microbiome plays an important role in soil and ecosystem health, including both beneficial members that provide key ecosystem goods and services as well as pathogens that jeopardise food safety and security. They also suggested that plant diseases can be mitigated by manipulating resident microbes in situ utilising a variety of agronomic methods, such as low tillage, crop rotation, cover cropping, and organic mulching. Furthermore, the use of microbial inoculants has demonstrated potential in disease management. Advances in DNA sequencing technology have made it easier to research the plant microbiome, revealing intricate relationships between plants and their microbial populations (Pucker *et al.*, 2022).

The contributions of microorganisms in mitigating the effects of climate change

Microorganisms play a significant role in reducing the consequences of climate change, since they actively produce and consume major greenhouse gases such as carbon dioxide (CO_2), methane (CH_4), and nitrous oxide. Anthropogenic changes have given bacteria easier access to carbon and nitrogen, contributing to higher amounts of these gases. Microbes that produce and consume these gases live in a variety of environments, which influences greenhouse gas dynamics (Bardgett & van der Putten, 2014). Furthermore, several pathogenic bacteria adapt to climate change by expanding their territories via variables such as insect vectors, floods, or strong storms, possibly impacting hosts susceptible to heat or drought stress.

Microorganisms might help to mitigate the consequences of climate change in a variety of ways, including:

- **Soil ecosystems and greenhouse gas dynamics:** Microorganisms, particularly those found in soil ecosystems, actively degrade soil organic matter, making it accessible to crops and altering greenhouse gas production and consumption rates (Bardgett & van der Putten, 2014). Microbes fix an estimated 70-140 million tonnes of nitrogen per year worldwide, offering an alternative to manufactured fertilisers and contributing to ecosystem resilience.

- **Global fluxes of greenhouse gases:** Microbial activities have a considerable effect on global biogenic greenhouse gas fluxes, and their quick reaction to climate change may be seen.
- **Adaptation to climate change:** Plant-associated microbes with characteristics such as drought tolerance impact plant responses to abiotic stresses, assisting in the adaptation of agricultural operations to changing climate conditions (Bardgett & van der Putten, 2014).
- **Terrestrial microbial processes:** Managing terrestrial microbial activities has potential for mitigating climate change by lowering greenhouse gas emissions.
- **Microbial communities and feedback responses:** Microbial communities vary their structure and composition in response to climate change, helping to solve environmental problems through food cycling and functional genetic material stimulation (Bardgett & van der Putten, 2014).
- **Biogeochemical cycles:** The relationship between microbial communities and biogeochemical cycles is an efficient strategy for climate change mitigation because microorganisms use greenhouse gases as an energy source and integrate them into their cellular structures (Falkowski *et al.*, 2008).
- **Oxygen absorption:** Microorganisms assist in mitigating climate change by absorbing oxygen, a critical aspect given the importance of carbon dioxide in the environment (Cavicchioli *et al.*, 2019).
- **Nitrous oxide emission reduction:** Efforts to minimise nitrous oxide emissions involve microbiological strategies such as manipulating microbiomes in various habitats and combining microbiome-based information of nitrous oxide source and sinks. These include employing bacteria-based systems to remove nitrous oxide directly from sources such as automobile exhaust and power plants (Hu *et al.*, 2017).
- **Smart agriculture applications and innovative technologies:** Smart agriculture applications, which aim to reduce the agricultural industry's long-term influence on the environment, require an understanding of microbial ecology and plant-microbe interactions (Das *et al.*, 2019). Furthermore, innovative technologies, such as "living concrete" that uses photosynthetic cyanobacteria, provide ecologically benign solutions by collecting CO2 from the atmosphere. Cyanobacteria may cause carbonate precipitation in both aquatic and terrestrial

environments; hence, researchers are interested in using cyanobacteria for the biomineralization process in cementitious structures.

Role of microbes in food security

Microbe role is crucial in ensuring food security worldwide, particularly in the context of addressing water scarcity, a critical factor influencing agricultural productivity. These microbes play variety of indispensable roles in contributing to sustainable and resilient food production systems, as elucidated below:

- **Nutrient cycling and soil health:** Microorganisms play an important role in nutrient cycling in the soil, breaking down organic matter and releasing key nutrients, guaranteeing soil quality, increasing soil fertility, and providing crops with needed ingredients for development (Schröder *et al.,* 2016). Healthy soils with robust microbial communities provide a substantial contribution to higher agricultural output.
- **Biofertilizers and nitrogen fixation:** Certain microorganisms, such as nitrogen-fixing bacteria and mycorrhizal fungus, create symbiotic interactions with plants, facilitating nutrient uptake. These biofertilizers improve nutrient utilisation efficiency, lowering the need for synthetic fertilisers, which is critical in water-scarce areas (Sánchez-Navarro *et al*., 2020).
- **Water use efficiency:** Microorganisms contribute to plant water usage efficiency, with specific microbial inoculants strengthening plants' ability to absorb and use water more efficiently, reducing the impact of water shortage on agricultural yields (Bardgett & van der Putten, 2014).
- **Disease resistance:** Beneficial microbes helps to improve plant health and disease resistance by functioning as biopesticides, protecting crops from dangerous pathogens and reducing the need for chemical treatments (El-Saadony *et al.,* 2022). This provides a better food supply while reducing the environmental effect associated with traditional pest control approaches.
- **Biological control of pests:** Microorganisms, such as bacteria and fungus, work as biological pest control agents, minimising the need for water-intensive chemical pesticides and supporting environmentally friendly pest management techniques (Bonaterra *et al.,* 2022).
- **Efficient waste decomposition:** Microorganisms are essential for the breakdown of organic waste, especially agricultural leftovers. Efficient decomposition feeds the soil with organic matter and helps to save water by reducing biomass buildup, which can interfere with water availability (Raza *et al*., 2023).

- **Improved crop resilience**: Microbial inoculants, such as plant growth-promoting rhizobacteria (PGPR) and mycorrhizal fungi, improve crop resistance to environmental stresses, including water shortages, by stimulating root development and boosting overall plant tolerance to water stress (El-Saadony *et al*., 2022).
- **Bioremediation of contaminated water and soil:** Microorganisms play an important part in bioremediation, which uses microbial activity to clean and cleanse polluted water and soil (Dave & Das, 2021). This procedure is critical for preserving the safety and quality of water and soil resources in agriculture.

Leveraging microbial biotechnology for sustainable agriculture and climate mitigation

Microbial biotechnology offers a possible solution for tackling long-standing social and environmental challenges, encouraging sustainable agriculture, and reducing climate change. Microbial biotechnology has great promise as a technique for sustainable agriculture and climate mitigation. The following are some important points:

- **Sustainable Agriculture:** Microbial biotechnology can help with sustainable agriculture by enhancing plant-microbe interactions, smart crop farming, and biodiversity. Microbes can improve disease resistance and productivity, allowing crops to be cultivated without the use of chemical compounds (Tan *et al.,* 2022). They can also help with nutrient cycling, disease prevention, soil health, and plant resilience (Kumar et al., 2023).
- **Climate mitigation:** Soil bacteria play an important role in minimising the effects of climate change. They have the ability to remove carbon dioxide (CO2) from the atmosphere through processes like as carbon fixation and storage in organic materials. This microbial-driven carbon storage might transform climate-smart farming practices, resulting in increased production and environmental protection (Kumar *et al.,* 2023).
- **Biofertilizers, bioprotectants and biostimulants**: Beneficial microbial communities can supply biofertilizers, bioprotectants, and biostimulants, as well as help plants to cope with many forms of abiotic stress like drought, salinity, heat, flooding, etc.(Abdul *et al.,* 2022; Yadav *et al.,* 2021).

Climate change and food insecurity: A global challenge

Climate change and food hunger are two of the most severe global issues we are facing today. Climate change is affecting food production systems all around the world, making it more difficult to cultivate crops and rear cattle. This is resulting in food shortages, price increases, and greater hunger. Furthermore, tackling the difficulties of climate change and food insecurity necessitates a multidimensional strategy that involves understanding the links between climate change and food production, as well as developing novel solutions to reduce their consequences.

Current challenges posed by climate change and food insecurity

Climate change and food hunger are two interrelated global issues that are escalating. The number of people experiencing severe food insecurity has risen from 135 million in 2019 to 345 million in 82 countries by June 2022. This increase is the result of a variety of events, including conflicts, supply chain disruptions, and the ongoing economic effects from the COVID-19 epidemic, which pushed food prices to record highs. (World Bank, 2022; Kirubanandan, 2023)

Global warming is altering weather patterns, resulting in heat waves, torrential rains, and droughts. These factors are influencing food production and contributing to rising food commodity prices, forcing an extra 30 million people in low-income countries into food insecurity (World Bank, 2022).

Approximately 80% of the worldwide population most exposed to crop failures and famine as a result of climate change lives in Sub-Saharan Africa, South Asia, and Southeast Asia, where agricultural families are disproportionately poor and vulnerable. A catastrophic drought triggered by an El Nino weather pattern or climate change has the potential to force millions more people into poverty (World Bank, 2022).

Need for innovative solutions

Climate change's consequences on food security and productivity must be mitigated through innovative solutions. These initiatives should include investments in agricultural infrastructure, crop and food diversity, long-term food storage systems, and sustainable agriculture training for local farmers (Swinnen *et al.,* 2022).

Many modern technical breakthroughs, including as solar-powered irrigation pumps, cold storage, genome-editing technologies, and value chain automation, provide promise for lowering emissions while improving output. These solutions provide mutually beneficial prospects for addressing hunger and climate change. Furthermore, inclusive innovation has the ability to transform food systems and help to end world hunger. There is a $15.2 billion funding shortfall for food system innovation, which is critical for eliminating hunger, keeping emissions below 2°C, and cutting water consumption by 10%. The World Economic Forum and the UN Food and Agriculture Organisation have teamed to create a roadmap to help nations accelerate inclusive food system innovation (Whiting, 2022).

Challenges and limitations

Microbial biotechnology has immense potential for tackling key global concerns such as climate change mitigation and food security; nevertheless, its application is fraught with difficulties and constraints. The following discussion will dive into these problems while also offering insights into potential remedies to such limitations:

Challenges in microbial biotechnology for climate change mitigation

- **Scale-up challenges:** Scaling up microbial activities for large-scale climate change mitigation is a big hurdle. A potential answer to such an issue might be reached by integrated techniques that combine microbial biotechnology with yet-to-be discovered engineering technologies (Smith *et al.,* 2020).
- **Complexity of microbial community and interaction:** Understanding and managing diverse microbial ecosystems for precise climate interventions is difficult. Furthermore, the complex interactions between different microbial species, as well as their reactions to changing environmental circumstances, make it difficult to accurately forecast and manage microbial activities. A potential solution to this difficulty is to employ modern omics technologies and systems biology methodologies to gain a better knowledge of microbial community dynamics (Banerjee *et al.,* 2021).

Limitations of microbial biotechnology for food security

- **Intensity of resources:** Some microbial methods can be resource expensive, making them unsuitable for broad usage in agriculture. A potential solution to such limits might be to focus research efforts on creating resource-efficient microbial technology (Hartman *et al.,* 2017).
- **Regulatory and ethical concerns:** Stringent rules and public scepticism may hamper widespread use of genetically modified bacteria in agriculture, but a clear and inclusive regulatory structure, along with public participation, may ease acceptance of GMOs.

Moreover, implementing the following general recommendations may prove beneficial in addressing the challenges indicated above:

- **Interdisciplinary collaborations:** Collaborations among scientists with various specialties, such as microbiologists, engineers, biochemists, agriculturists, and environmental scientists, can help to create comprehensive solutions for scaling up microbial processes.
- **Innovation and research funding:** Governments and the corporate sector should invest in creative research financing structures that support long-term, high-risk microbial biotechnology initiatives addressing climate change and food security.
- **Education and outreach:** Increased efforts in public involvement and education can alleviate fears and increase acceptance of microbial biotechnologies for a variety of valuable uses. Regarding climate change and food security.

While microbial biotechnology offers promising prospects for combating climate change and food security, overcoming the problems needs collaborative, interdisciplinary efforts, novel financing structures, and open communication with the public. Continuous research and technical improvements will be critical in overcoming these obstacles and realising the full potential of microbial biotechnology in sustainable agriculture and climate change mitigation.

Microbial biotechnology is a lively subject with exciting opportunities for future study and growth. Recent advances highlight the promise in a variety of fields, including synthetic biology, environmental applications, and therapeutic treatments. Exploration of engineered microbial systems using synthetic biology tools for sustainable bio-manufacturing (Rojo et al., 2023), harnessing microbial consortia for environmental remediation, and innovative CRISPR-based tools for precision genome editing in microbes (Wei & Li, 2023) are just a few examples. These recent breakthroughs highlight the various and multidimensional prospects that await in the changing environment of microbial biotechnology.

Future perspectives

Microbial biotechnology has significant potential for tackling climate change and food hunger, providing novel solutions that harness the power of microorganisms to address these global issues. The interaction of microbial activities with the environment offers rich ground for study and development, with applications ranging from sustainable agriculture to bioenergy generation.

The following potential research areas in microbial biotechnology are promising for tackling climate change and food security:

- **Synthetic microbial communities:** Learning more about how they are assembled and function might help them perform better in agriculture and the environment.
- **Omics technologies:** Advances in metagenomics, Meta transcriptomics, and metabolomics will allow for a better knowledge of microbial populations and interactions.
- **Microbial-plant signalling**: Investigating the potential of extremophile microorganisms in severe environments might lead to new opportunities for sustainable agriculture in difficult areas.
- **Microbial engineering for extreme environments:** Investigating the potential of extremophile microorganisms in hard climates may offer up new options for sustainable agriculture in difficult areas.

Conclusion

This chapter highlights microorganisms' amazing resilience and effect on environmental health. Their critical involvement in soil health, nutrient cycling, and plant interactions lays the groundwork for climate resistance. Furthermore, emerging microbial technologies such as biofertilizers and biopesticides provide a sustainable alternative to traditional agricultural techniques, opening the path for lower greenhouse gas emissions and higher food production.

We must form symbiotic relationships with bacteria. By incorporating innovative microbial solutions into agriculture, we can adjust to changing weather patterns, handle water scarcity, and increase production in both traditional and precision agricultural systems. This review highlights the need of embracing microbial power. Doing so offers a future in which we cohabit with these little allies, cultivating a resilient and resource-efficient agricultural environment that ensures food security for years to come.

Further research should focus on unlocking the full potential of previously unknown microbial strains for climate mitigation and soil improvement, optimising the application of microbial technologies for diverse agricultural ecosystems and environmental conditions, and addressing potential social and economic challenges associated with the widespread adoption of microbial solutions.

As scientists continue to discover the secrets of the microbial world, we are on the verge of a revolution in agriculture and environmental sustainability. The future rests in cultivating this symbiotic alliance, and the benefits—a healthy world and secure food for all—are well worth the effort.

References

Abdul, K., Perveen, S., Alamer, K. H., Haq, M. Z. Ul, Rafique, Z., Alsu- days, I. M. et al. (2022). Arbuscular mycorrhizal fungi symbiosis to enhance plant-soil interaction. Sustainability, 14(13), 7840. https:// doi.org/10.3390/su14137840.

Affoh, Raïfatou, Zheng, H., Dangui, K., & Dissani, B. M. (2022). The impact of climate variability and change on food security in sub- saharan africa: Perspective from panel data analysis. Sustainability, 14(2), 759. https://doi.org/10.3390/su14020759.

Aguilar-Paredes, A., Valdés, G., Araneda, N., Valdebenito, E., Hansen, F., & Nuti, M. (2023). Microbial community in the composting process and its positive impact on the soil biota in sustain- able agriculture. Agronomy, 13(2), 542. https://doi.org/10.3390/agronomy13020542.

Ahuja, D. (2009). Sustainable energy for developing countries. Sapiens, 10(2), 105–120. http://journals.openedition.org/sapiens/823.

Anthony, M. A., Bender, F. S., & van der Heijden, M. G. A. (2023). Enumerating soil biodiversity. Proceedings of the National Academy of Sciences, 120, e2304663120.

Arrigo, K. R. (2005). Marine microorganisms and global nutrient cycles. Nature, 437(7057), 349–355.

Bailey-Serres, J., Parker, J. E., Ainsworth, E. A., Oldroyd, G. E. D., & Schroeder, J. I. (2019). Genetic strategies for improving crop yields. Nature, 575(7781), 109–118. https://doi.org/10.1038/ s41586- 019- 1679- 0

Banerjee, S., Schlaeppi, K., & van der Heijden, M. G. A., (2021). Keystone taxa as drivers of microbiome structure and functioning. Nature Reviews Microbiology, 19(9), 567–580.

Bardgett, R., & van der Putten, W., (2014). Below ground biodiversity and ecosystem functioning. Nature, 515, 505–511. https://doi. org/10.1038/nature13855.

Bashan, Y., de-Bashan, L. E., Prabhu, S. R., & Hernandez, J. P. (2016). Advances in plant growth-promoting bacterial inoculant technology: Formulations and practical perspectives (1998–2013). Plant and Soil, 378(1–2), 1–33.

Bhardwaj, D., Ansari, M. W., Sahoo, R. K., Tuteja, N. (2014). Biofertilizers function as key player in sustainable agriculture by improving soil fertility, plant tolerance and crop productivity. Microbial Cell Factories, 13, 66. https://doi.org/10.1186/1475-2859-13-66.

Bhattacharyya, P. N., Jha, D. K., & Plant, A. B. (2020). Plant growth- promoting rhizobacteria (PGPR): Emergence in agriculture. World Journal of Microbiology and Biotechnology, 36, 191.

Bolhuis, H., & Cretoiu, M. S. (2016). What is so special about marine microorganisms? In L. J. Stal, M. S. Cretoiu (Eds.), The Marine Microbiome (pp. 3–20). Springer.

Bonaterra, A., Badosa, E., Daranas, N., Francés, J., Roselló, G., & Montesinos, E. (2022). Bacteria as biological control agents of plant diseases. Microorganisms, 10(9), 1759. https://doi.org/10.3390/microorganisms10091759.

Brzozowski, L., & Mazourek, M. (2018). A sustainable agricultural future relies on the transition to organic agroecological pest management. Sustainability, 10(6), 2023. https://doi.org/10.3390/su10062023.

Cavicchioli, R., Ripple, W. J., Timmis, K. N., Azam, F., Bakken, L. R., Baylis, M. et al. (2019). Scientists' warning to humanity: Microorganisms and climate change. Nature Reviews Microbiology, 17, 569–586. https://doi.org/10.1038/s41579-019-0222-5.

Chakraborty, A., & Kundu, S. (2023). Chapter-8 symbiotic nitrogen fixation and soil fertility chapter-8 symbiotic nitrogen fixation and soil fertility. https://www.researchgate.net/publication/373632503_Chapter.

Chen, W., Modi, D., & Picot, A. (2023). Soil and phytomicrobiome for plant disease suppression and management under climate change: A review. Plants, 12(14), 2736. https://doi.org/10.3390/ plants12142736.

Cheng, F., & Cheng, Z. (2015). Research progress on the use of plant allelopathy in agriculture and the physiological and ecological mechanisms of allelopathy. Frontiers in Plant Science, 6. https://doi. org/10.3389/fpls.2015.01020.

Compant, S., Duffy, B., Nowak, J., Clement, C., & Barka, E. A. (2021). Use of plant growth-promoting bacteria for biocontrol of plant diseases: Principles, mechanisms of action, and future prospects. Applied and Environmental Microbiology, 87(4), e00547–20.

Damalas, C. A., & Eleftherohorinos, I. G. (2011 May). Pesticide exposure, safety issues, and risk assessment indicators. International Journal of Environmental Research and Public Health, 8(5), 1402– 1419. https://doi.org/10.3390/ijerph8051402.

Das, S., Ho, A., & Pil, K. M. (2019). Editorial: Role of microbes in climate smart agriculture. Frontiers in Microbiology, 10. https://doi. org/10.3389/fmicb.2019.02756.

Das, S., Ray, M. K., Panday, D., & Mishra, P. K. (2023). Role of biotechnology in creating sustainable agriculture. PLOS Sustainability and Transformation, 2(7), e0000069. https://doi.org/10.1371/journal.pstr.0000069.

Dave, S., & Das, J. (2021). Role of microbial enzymes for biodegradation and bioremediation of environmental pollutants: Challenges and future prospects. Bioremediation for Environmental Sustainability, 325–346. https://doi.org/10.1016/B978-0-12-820524-2.00013-4.

Delgado-Baquerizo, M., Oliverio, A. M., Brewer, T. E., Benavent- González, A., Eldridge, D. J., Bardgett, R. D. et al. (2018). A global atlas of the dominant bacteria found in soil. Science, 359(6373), 320–325. https://doi.org/10.1126/science.aap9516.

De Sousa, CS., Hassan, SS., Pinto, AC., Silva, WM., De Almeida, SS., De Castro, SS. et al. (2018). Chapter 1—microbial omics: Applications in biotechnology. In D. Barh & V. Azevedo (Eds.), Omics Technologies and Bio-Engineering. pp. (3–20). Cambridge: Academic Press.

Dhanaraju, M., Chenniappan, P., Ramalingam, K., Pazhanivelan, S., & Kaliaperumal, R. (2022). Smart farming: Internet of things (IoT)- based sustainable agriculture. Agriculture, 12(10), 1745. https://doi.org/10.3390/agriculture12101745.

Dong, O. X., & Ronald, P. C. (2019). Genetic engineering for disease resistance in plants: Recent progress and future perspectives. Plant Physiology, 180(1), 26–38. https://doi.org/10.1104/pp.18.01224.

El-Saadony, M. T., Saad, A. M., Soliman, S. M., Salem, H. M., Ahmed, A. I., Mahmood, M. et al. (2022). Plant growth-promoting microorganisms as biocontrol agents of plant diseases: Mechanisms, challenges and future perspectives. Frontiers in Plant Science, 13, 923880. https://doi.org/10.3389/fpls.2022.923880.

EPA (2023). Climate change impacts on agriculture and food supply. Retrieved Nov 22. https://www.epa.gov/climateimpacts/climate- change- impacts- agriculture- and- food- supply.

Falkowski, P. G., Fenchel, T., & Delong, E. F. (2008). The microbial engines that drive Earth's biogeochemical cycles. Science, 320(5879), 1034–1039. https://doi.org/10.1126/science.1153213.

FAO (2017). Marine Protected Areas: Interactions with Fishery Livelihoods and Food Security. FAO Fisheries and Aquaculture Technical Paper 603. Rom: Food and Agriculture

Organization of the United Nations and International Union for Conservation of Nature. https://www.fao.org/3/i6742e/i6742e.pdf .

Fierer, N. (2017). Embracing the unknown: Disentangling the complexities of the soil microbiome. Nature Reviews Microbiology, 15(10), 579–590.

Glöckner, F. O., Gasol, J. M., McDonough, N., Calewaert, J. B. et al. (2012). Marine microbial diversity and its role in ecosystem functioning and environmental change. European Science Foundation, Position paper 17. ISBN 978-2-918428-71-8.

Gänzle, M. G. (2015). Lactic metabolism revisited: Metabolism of lactic acid bacteria in food fermentations and food spoilage. Current Opinion in Food Science, 2, 106–117.

Hartman, K., van der Heijden, M. G. A., Wittwer, R. A., Banerjee, S., & Walser, J. C. (2017). Cropping practices manipulate abundance patterns of root and soil microbiome members paving the way to smart farming. Microbiome, 5(1), 14.

Herrero, M., Thornton, P. K., Mason-D'Croz, D., Palmer, J., Bodirsky, B. L., Pradhan, P. et al. (2021). Articulating the effect of food systems innovation on the sustainable development goals. The Lancet Planetary Health, 5(1), e50–e62. https://doi.org/10.1016/ S2542-5196(20)30277-1.

Hu, H. W., He, J. Z., & Singh, B. (2017). Harnessing microbiome- based biotechnologies for sustainable nitrous oxide emissions. Microbial Biotechnology, 10(5), 1226–1231. https://doi. org/10.1111/1751- 7915.12758.

Hussein, E. I., Al-Horani, F. A., & Malkawi, H. I. (2012). Bioremediation capabilities of oil-degrading bacterial consortia isolated from oil- contaminated sites at the Gulf of Aqaba (Jordan). Biotechnology, 11(4), 189–198. https://doi.org/10.3923/biotech.2012.189.198.

Ibáñez, A., Garrido-Chamorro, S., & Barreiro, C. (2023). Microorganisms and climate change: A not so invisible effect. Microbiology Research, 14(3), 918–947. https://doi.org/10.3390/ microbiolres14030064.

IMF (2022). Climate change and chronic food insecurity in sub- Saharan Africa. International Monetary Fund, WP/22/16. https://www.imf.org/en/Publications/Departmental-Papers-Policy -Papers/Issues/2022/09/13/Climate-Change-and-Chronic-Food-In security-in-Sub-Saharan-Africa-522211.

International Energy Agency (IEA) (2021). Sources of greenhouse gas emissions. https://www.epa.gov/ghgemissions/sources-greenhouse-gas-emissions.

International Food Policy Research Institute (IFPRI) (2022). Global food policy report: Climate change and food systems. Washington, DC: International Food Policy Research Institute (IFPRI). https://doi.org/10.2499/9780896294257.

International Panel on Climate Change (IPCC) (2022). Version 13. https://www.ipcc.ch/report/ar6/wg1/downloads/outreach/IPCC_AR6_WGI_SummaryForAll.pdf].

IPCC (2019). Summary for policymakers. In P. R. Shukla, J. Skea, E.C. Buendia, V. Masson-Delmotte, H.-O. Pörtner, D. C. Roberts et al. (Eds.), Climate Change and Land: An IPCC Special Report on Climate Change, Desertification, Land Degradation, Sustainable Land Management, Food Security, and Greenhouse Gas Fluxes in Terrestrial Ecosystems. Chapter 5: Food Security.

Isaacs-Thomas, B. (Feb 28, 2022). How climate change is hurting living things on Earth right now, according to a new report. https://www.pbs.org/newshour/science/3-things-to-know-about-climate-change-hurting-earths-inhabitants-and-how-to-deal-with-it.

Javed, A., Ali, E., Afzal, K. B., Osman, A., & Riaz, S. (2022). Soil fertility: Factors affecting soil fertility, and biodiversity responsible for soil fertility. International Journal of Plant, Animal and Environmental Sciences, 12, 021–033. https://doi.org/10.26502/ ijpaes.202129.

Jiao, N., Liu, J., Jiao, F., Chen, Q., & Wang, X. (2020). Microbes mediated comprehensive carbon sequestration for negative emissions in the ocean. National Science Review, 7(12), 1858–1860. https://doi. org/10.1093/nsr/nwaa171.

Jones, M. W., Peters, G. P., Gasser, T., Andrew, R.M., Schwingshackl, C., Gütschow, J. et al. (2023). National contributions to climate change due to historical emissions of carbon dioxide, methane, and nitrous oxide since. 1850. Scientific Data, 10, 155. https://doi. org/10.1038/ s41597-023-02041-1.

Kirubanandan, S. (2023). How to mitigate the effects of climate change on global food security. Apr 6, 2023. World Eco- nomic Forum. https://www.weforum.org/agenda/2023/04/ mitigate- climate- change- food- security.

Kougias, P. G., Kim, J., Shen, T., & Angelidaki, I. (2021). Anaerobic digestion for sustainable energy production and nutrient recovery from wastewater. Trends in Environmental Analytical Chemistry, 30, e00115.

Koza, N. A., Adedayo, A. A., Babalola, O. O., & Kappo, A. P. (2022). Microorganisms in plant growth and development: Roles in abiotic stress tolerance and secondary metabolites secretion. Microorganisms, 10(8), 1528. https://doi.org/10.3390/ microorganisms10081528.

Kumar, J. D., Prakash, V. J., Tarun, B., Prudêncio De Araujo, P. A., & Bapurao, A. A. (2023). Editorial: Microbial co-cultures: A new era of synthetic biology and metabolic engineering. Frontiers in Microbiology, 14. https://doi.org/10.3389/fmicb.2023.1235565.

Kumar, K., Gambhir, G., Dass, A., Tripathi, A.K., Singh, A., Jha, A.K., Yadava, P., Choudhary, M., Rakshit, S. (2020). Genetically modified crops: Current status and future prospects. Planta, 251(4), 91. https://doi.org/10.1007/s00425-020-03372-8. PMID: 32236850.

Kumar, P., Mishra, S., & Malik, A. (2021). Integrated pest management: A component of sustainable agriculture. In Microbial Biotechnology for Sustainable Agriculture (pp. 189–212). Springer.

Kumari, A., Dash, M., Singh, S. K., Jagadesh, M., Mathpal, B., Mishra, P. K. et al. (2023). Soil microbes: A natural solution for mitigating the impact of climate change. Environmental Monitoring and Assessment, 195, 1436. https://doi.org/10.1007/s10661-023-11988-y.

Kuypers et al. (2018). The microbial nitrogen-cycling network. Nature reviews microbiology. https://www.nature.com/articles/ nrmicro.2018.9.

Kyriakopoulos, G. L. (2023). Enhancing climate neutrality and resilience through energy efficiency policies in households: A systematic literature review. Sustainability, 11(5), 105. https://www.mdpi. com/2225- 1154/11/5/105.

Lake, I. R., & Barker, T. (2012). Climate change and food security: Health impacts in developed countries. Emerging Themes Epidemiology, 7(1), 1–9. https://www.ncbi.nlm.nih.gov/ pmc/articles/ PMC3556605/.

licón-Hernández, D. R., Guerra-Sánchez, G., Porta, C. J., Santoyo- Tepole, F., Hernández-Cortez, C., Tapia-García, E. Y. et al. (2022). Fundaments and concepts on screening of microorganisms for biotechnological applications. Mini review. Current Microbiology, 79(12), 373. https://doi.org/10.1007/s00284-022-03082-2.

Lindwall, C. (October 24, 2022). What are the effects of climate change? NRSDC. https://www. nrdc.org/stories/what- are- effects- climate- change.

Lopes, M. S., & da Silva, G. P., (2021). Microbial biofuel production: Recent advances and future perspectives. Frontiers in Bioengineering and Biotechnology, 9, 637405.

Louca, S., Parfrey, L. W., & Doebeli, M. (2016). Decoupling function and taxonomy in the global ocean microbiome. Science, 353(6305), 1272–1277.

Luo, G., Xue, C., Jiang, Q., Xiao, Y., Zhang, F., Guo, S. et al. (2020). Soil carbon, nitrogen, and phosphorus cycling microbial populations and their resistance to global change depend on soil C:N:P stoi- chiometry. mSystems J30, 5(3), e00162–20. https://doi.org/10.1128/mSystems.00162- 20.

Malkawi, H. I., & Al-Omari, M. N. (2010). Culture-dependent and culture-independent approaches to study the bacterial and Archaeal diversity from jordanian hot springs. African Journal of Microbiology Research, 4(10), 923–932.

Malkawi, H. I., & Kapiel, T. Y. S. (2024). Microbial Biotechnology: A Key Tool for Addressing Climate Change and Food Insecurity. European Journal of Biology and Biotechnology, 5(2), 1-15.

Marinina, O., Nechitailo, A., Stroykov, G., Tsvetkova, A., Reshneva, E., & Turovskaya, L. (2023). Technical and economic assessment of energy efficiency of electrification of hydrocarbon production facilities in underdeveloped areas. Sustainability, 15(12), 9614. https:// doi.org/10.3390/su15129614.

Mayer, M. (2018). What are environmental problems due to population growth? Sciencing.com. https://sciencing.com/environmental- problems- due- population- growth- 8337820.html.

Mohammad, M. J., Malkawi, H. I., & Shibli, R. (2003). Effects of arbuscular mycorrhizal fungi and P fertilization on growth and nutrition uptake of barley grown on soils with different levels of salts. Journal of Plant Nutrition, 26(1), 125–137.

Mosunova, O., Navarro-Muñoz, J. C., & Collemare, J. (2021). The biosynthesis of fungal secondary metabolites: From fundamentals to biotechnological applications. In Z. Óscar, C. Arturo (Eds.), Encyclopedia of Mycology (pp. 458–476). Elsevier. https:// doi.org/10.1016/B978-0-12-809633-8.21072-8.

Myers, S. S. (2022). Food security, climate change, and health. Journal of the Royal Society of Medicine, 115(2), 73–81. https://www.ncbi.nlm. nih.gov/pmc/articles/PMC9517947/.

Nadeem, S. M., Ahmad, M., & Zahir, Z. A. (2021). Plant growth- promoting microorganisms and sustainable agriculture: An overview. In Microbial Biotechnology for Sustainable Agriculture (pp. 1–18). Springer.

Nazaries, L., Pan, Y., Bodrossy, L., Baggs, E. M., Millard, P., Murrell, J. C. et al. (2021). Evidence for microbial 'Methane Sink' activity in soils from methane and ammonium enrichment. Environmental Microbiology, 23(1), 575–586.

Ondik, M. M., Ooi, M. K. J., & Muñoz-Rojas, M. (2023). Soil microbial community composition and functions are disrupted by fire and land use in a Mediterranean woodland. Science of the Total Environ- ment, 895, 165088. https://doi.org/10.1016/j.scitotenv.2023.165088.

Onen, I. O., Aboh, A. A., Mfam, A. N., Akor, M. O., Nweke, C. N., & Osuagwu, A. N. (2020). Microbial diversity: Values and roles in ecosystems. Asian Journal of Biology, 9(1), 10–22. https://doi. org/10.9734/ajob/2020/v9i130075.

Onyeaka, H., & Ekwebelem, O. C. (2023). A review of recent advances in engineering bacteria for enhanced CO2 capture and utilization. International Journal of Environmental Science and Technology, 20, 4635–4648. https://doi.org/10.1007/s13762-022-04303-8.

Panda (2023). A growing need for species to adapt to a changing world. Retrieved Nov 22, 2023. https://wwf.panda.org/discover/our_focus/wildlife_practice/problems/climate_change .

Poblete-Castro, I., Wittmann, C., & Nikel, P. I. (2020). Biochemistry, genetics and biotechnology of glycerol utilization in Pseudomonas species. Microbial Biotechnology, 13, 32–53. https://doi. org/10.1111/1751- 7915.13400.

Pucker, B., Irisarri, I., De Vries, J., & Xu, B. (2022). Plant genome sequence assembly in the era of long reads: Progress, challenges and future directions. Quantitative Plant Biology, 3, E5. https://doi. org/10.1017/qpb.2021.18.

Raveendran, S., Parameswaran, B., Ummalyma, S. B., Abraham, A., Mathew, A. K., Madhavan, A. et al. (2018). Applications of microbial enzymes in food industry. Food Technology and Biotechnology, 56(1), 16–30. https://doi.org/10.17113/ftb.56.01.18.5491.

Raza, T., Qadir, MF., Khan, KS., Eash, NS., Yousuf, M., Chatterjee, S. et al. (2023). Unrevealing the potential of microbes in decomposition of organic matter and release of carbon in the ecosystem. Journal of Environmental Management, 344, 118529. https://doi.org/10.1016/j. jenvman.2023.118529.

Reisoglu, S ̦., & Aydin, S. (2023). Microalgae as a promising candidate for fighting climate change and biodiversity loss. Microalgae-current and potential applications. IntechOpen. https://doi.org/10.5772/ intechopen.1002414.

Rojo, F. P., Vuong, P., & Pillow, J. J. (2023). Microbial synthetic biology for plant metabolite production: A strategy to reconcile human health with the realization of the UN sustainable development goals. Biofpr, 17(6), 1485–1495. https://doi.org/10.1002/bbb.2522.

Schröder, JJ., Schulte R.P., O., Creamer, RE., Delgado, A., van Leeuwen, J., Lehtinen, T. et al. (2016). The elusive role of soil quality in nutrient cycling: A review. Soil Use and Management, 32(4), 476– 486. https://doi.org/10.1111/sum.12288.

Sidhu, N., Goyal, S., & Reddy, M. S. (2022). Biomineralization of cyanobacteria Synechocystis pevalekii improves the durability properties of cement mortar. AMB Express, 12(1), 59. https://doi. org/10.1186/s13568- 022- 01403- z.

Singh, B. K., Delgado-Baquerizo, M., Egidi, E., Guirado, E., Leach, J. E., Liu, H. et al. (2023). Climate change impacts on plant pathogens, food security and paths forward. Nature Reviews Microbiology, 21, 640–656. https://doi.org/10.1038/s41579-023-00900-7.

Six, J., Bossuyt, H., Degryze, S., & Denef, K. (2004). A history of research on the link between (micro) aggregates, soil biota, and soil organic matter dynamics. Soil and Tillage Research, 79(1), 7–31.

Smith, M. A., Lam, M. M., & Nielsen, L. K. (2020). Overcoming the challenges of scaling up microbial processes. Trends in Biotechnology, 38(6), 586–599.

Smith, N. W., Fletcher, AJ., Millard, P., Hill, JP., & McNabb, WC. (2022). Estimating cropland requirements for global food system scenario modeling. Frontiers in Sustainable Food Systems, 6. https://doi.org/10.3389/fsufs.2022.1063419.

Smith, S. E., & Read, D. (2010). Mycorrhizal Symbiosis. Academic Press. Sposito, G. (2023). Soil. In Encyclopedia Britannica. https://www.britannica.com/science/soil.

Swinnen, J., Arndt, C., & Vos, R. (12 May 2022). IFPRI global food policy report 2022: Accelerating food systems transformation to combat climate change. https://reliefweb.int/report/world/2022- global- food- policy- report-climate- change- and- food- systems.

Sánchez-Navarro, V., Zornoza, R., Faz, Á, Egea-Gilabert, C., Ros, M., Pascual, J. A. et al. (2020). Inoculation with different nitrogen-fixing bacteria and arbuscular mycorrhiza affects grain protein content and nodule bacterial communities of a fava bean crop. Agronomy, 10(6), 768. https://doi.org/10.3390/agronomy10060768.

Tan, C., Kalhoro, M. T., Faqir, Y., Ma, J., Osei, M. D., & Khaliq, G. et al. (2022). Climate-resilient microbial biotechnology: A perspective on sustainable agriculture. Sustainability, 14(9), 5574. https://doi.org/10.3390/su14095574.

Tedersoo, L., Bahram, M., Põlme, S., Kõljalg, U., Yorou, N. S., Wijesun- dera, R. et al. (2014). Global diversity and geography of soil fungi. Science, 346(6213), 1256688.

Thakur, N., Singh, S. P., & Zhang, C. (2022). Microorganisms under extreme environments and their applications. Current Research in Microbial Sciences, 3, 100141. https://doi.org/10.1016/j. crmicr.2022.100141.

The White house (2022). Executive order on advancing biotechnology and biomanufacturing innovation for a sustainable, safe, and secure American bioeconomy. https://www.whitehouse.gov/ briefing-room/presidential-actions/2022/09/12/executive-order-on-advancing- biotechnology- and- biomanufacturing- innovation- for- a- sustainable- safe-and- secure- american- bioeconomy.

Trump, B., Cummings, C., Klasa, K., Galaitsi, S., & Linkov, I. (2023). Governing biotechnology to provide safety and security and address ethical, legal, and social implications. Frontiers in Genetics, 13, 1052371. https://doi.org/10.3389/fgene.2022.1052371.

UN-Habitat (2023). Cities and citizens series: Cairo—a city in transition. https://issuu.com/unhabitat/docs/cities_and_citizen_series-_ bridging_the_urban_divi.

USGS (2023). How does climate change affect the challenge of invasive species? USGS. Retrieved Nov 22, 2023. https://www.usgs.gov/faqs/ how- does- climate- change- affect-challenge- invasive- species.

Victor, O., Chiza, K., Beatrice, E., Parker Megan, E., Laina, E., Nanna, R. et al. (2022). The impact of climate change on food systems, diet quality, nutrition, and health outcomes: A narrative review. Frontiers in Climate, 4. https://doi.org/10.3389/fclim.2022.941842.

Wang, G., Ren, Y., Bai, X., Su, Y., & Han, J. (2022 Nov 23). Contributions of beneficial microorganisms in soil remediation and quality improvement of medicinal plants. Plants, 11(23), 3200. https://doi.org/10.3390/plants11233200.

Watts, N., Amann, M., Arnell, N., Ayeb-Karlsson, S., Beagley, J., Belesova, K. et al. (2021 Jan 9). The 2020 report of the Lancet countdown on health and climate change: Responding to converging crises. Lancet, 397(10269), 129–170. https://doi.org/10.1016/ S0140-6736(20)32290- X.

Wei, J., & Li, Y. (2023). CRISPR-based gene editing technology and its application in microbial engineering. Engineering Microbiology, 3(4), 100101. https://doi.org/10.1016/j.engmic.2023.100101.otics.

Whiting, K. (2022). How inclusive innovation could transform food systems—and help to end world hunger. Food security, world economic forum. Retrieved Nov. 26, 2023. https://www.weforum.org/ agenda/2022/03/food-systems-innovation-transformation.

World Bank (2022). What you need to know about food security and climate change. Climate explainer. Retrieved Nov. 26, 2023.https:// www.worldbank.org/en/news/feature/2022/10/17/what- you- need- to- know- about- food-security- and- climate-change.

World Meteorological Organization (WMO) (2022). WMO provisional state of the global climate 2022. https://reliefweb.int/report/world/wmo-provisional- state- global- climate- 2022.

Yadav, A. N., Singh, J., Singh, C., & Yadav, N. Editors (2021). In Book: Current Trends in Microbial Biotechnology for Sustainable Agriculture. Singapore: Springer.

Yafetto, L. (2020). Application of solid-state fermentation by microbial biotechnology for bioprocessing of agro-industrial wastes from 1970 to: A review and bibliometric analysis. Heliyon, 8(3), e09173. https://doi.org/10.1016/j.heliyon.2022.e09173.

Zhang, Q., Zhang, Z., & Wang, D. (2020). The contribution of soil microbial diversity to ecosystem multifunctionality. Soil Ecology Letters, 2(1), 63–72.

9

Mitigation and Adaptation Strategies for Climate Change and Microbiology

Abarna Ravichandran and Jinisha Blessie J.P.

Department of Agricultural Microbiology
College of Agriculture, Vellayani, KAU, Thiruvananthapuram, Kerala

Abstract

Climate change, driven primarily by human activities such as the burning of fossil fuels and deforestation, has emerged as one of the most pressing global challenges. This complex phenomenon has far-reaching implications for various ecosystems, including the intricate relationships between climate and microbial communities. Microorganisms, including bacteria, archaea, fungi, and viruses, play pivotal roles in maintaining ecosystem stability and functioning. Climate change-induced alterations in temperature, precipitation patterns, and extreme weather events directly influence microbial communities. Shifts in microbial diversity and composition have been observed in diverse ecosystems, ranging from soil and oceans to the human microbiome. These changes can have cascading effects on nutrient cycling, carbon sequestration, and overall ecosystem resilience. Microbes are integral components of these cycles, influencing greenhouse gas emissions, nutrient availability, and soil fertility. Understanding how climate change affects microbial processes is essential for predicting ecosystem responses and developing strategies for mitigating the associated environmental consequences. Microbes, however, are not passive bystanders in the face of climate change; they possess remarkable adaptive capacities. Some microbial species demonstrate resilience to altered environmental conditions, while others may undergo evolutionary changes to thrive in new niches. The intricate interplay between climate change and microbiology is a critical aspect of understanding and addressing the broader challenges posed by environmental transformations.

Keywords: *Microorganisms, climate change, nutrient cycling, carbon sequestration*

Introduction

Climate change is widely acknowledged as the most significant contemporary challenge facing humanity. According to a recent report from the Intergovernmental Panel on Climate Change (IPCC, 2022), the situation has deteriorated further, with 3.3 billion people worldwide highly susceptible to climate change. Current unsustainable development practices are escalating the exposure of ecosystems and populations to climate-related risks. Human activities and their impact on the climate and environment are leading to unprecedented extinctions of animals and plants, resulting in a loss of biodiversity and posing a threat to life on Earth. The diversity of microorganisms is fundamental to maintaining a healthy global ecosystem, as the microbial world serves as the life support system of the biosphere. Although the influence of human activities on microorganisms is less conspicuous and less well-understood, a significant concern is that changes in microbial biodiversity and functions will impact the resilience of other organisms, affecting their ability to adapt to climate change. Microorganisms play crucial roles in carbon and nutrient cycling, as well as in the health of animals (including humans), plants, agriculture, and the global food web. Microorganisms inhabit all environments on Earth occupied by larger organisms and are the exclusive life forms in certain environments such as the deep subsurface and extreme habitats. Dating back to the origin of life on Earth at least 3.8 billion years ago, microorganisms are likely to persist well beyond any future extinction events. Despite their vital role in regulating climate change, microorganisms are seldom the focal point of climate change studies and are typically overlooked in policy development. The challenge lies in understanding their intricate role in the ecosystem due to their vast diversity and varied responses to environmental changes.

Climate change affects microorganisms

Changes in climate can impact the composition and diversity of microbial communities both directly, such as factors like seasonality and temperature, and indirectly, influenced by factors like plant composition, plant litter, and root exudates. The diversity of soil microorganisms is crucial for ecosystem functions, particularly carbon cycling, and has a significant influence on plant diversity. Experiments involving short-term laboratory warming and natural geothermal warming over more than 50 years have shown that initial increases in the growth and respiration of soil microorganisms led to a net release of CO_2 and subsequent substrate depletion. This resulted in decreased biomass and reduced microbial activity, suggesting that microbial communities do not readily acclimate to higher temperatures. However, a decade-long study demonstrated that soil communities can adapt to increased temperatures by altering composition and substrate use patterns, ultimately leading to less carbon loss compared to non-adapted conditions. In a 26-year forest-soil warming study, changes in organic matter decomposition and CO_2 release occurred over time, influencing microbial community composition and carbon use efficiency. This resulted in reduced microbial biomass and decreased availability of microbially accessible carbon. Climate change impacts microbial communities directly and indirectly through various interconnected factors, including temperature, precipitation, soil properties, and plant input. In desert environments where soil microorganisms are carbon-limited, increased carbon input from plants promotes the transformation of nitrogenous compounds, microbial biomass, diversity (such as fungi), enzymatic activity, and the utilization of recalcitrant organic matter. While these changes may enhance respiration and overall carbon loss from the soil, the unique characteristics of arid and semiarid regions suggest they could potentially function as carbon sinks. Additionally, climate change is anticipated to escalate the frequency, intensity, and duration of cyanobacterial blooms in eutrophic lakes, reservoirs, and estuaries. Cyanobacteria that form these blooms produce various toxins, posing threats to birds, mammals (including waterfowl, cattle, and dogs), and the usability of water bodies for recreation, drinking water production, agricultural irrigation, and fisheries. Notable instances of toxic cyanobacteria causing water quality issues include Lake Taihu (China), Lake Erie (USA), Lake Okeechobee (USA), Lake Victoria (Africa), and the Baltic Sea.

Farming and various human practices that impact microorganisms

Agricultural activities exert specific influences on microbial communities. Factors such as land use, including the types of plants grown, and sources of pollution, such as fertilizers, disrupt the composition and function of microbial communities, leading to changes in the natural cycles of carbon, nitrogen, and phosphorus transformations. Methanogens, found in environments like saturated soils with anaerobic conditions (e.g., rice paddies and constructed wetlands) and directly associated with ruminant animals (e.g., cattle, sheep, and goats), contribute significantly to methane production are illustrated in Fig. 1 (Galfalk *et al.*, 2016). Human actions that result in a decrease in microbial diversity also diminish the ability of microorganisms to support plant growth. It is imperative to gain insights into the impacts of climate change and other human activities on the microbial processes involved in transforming nitrogen compounds.

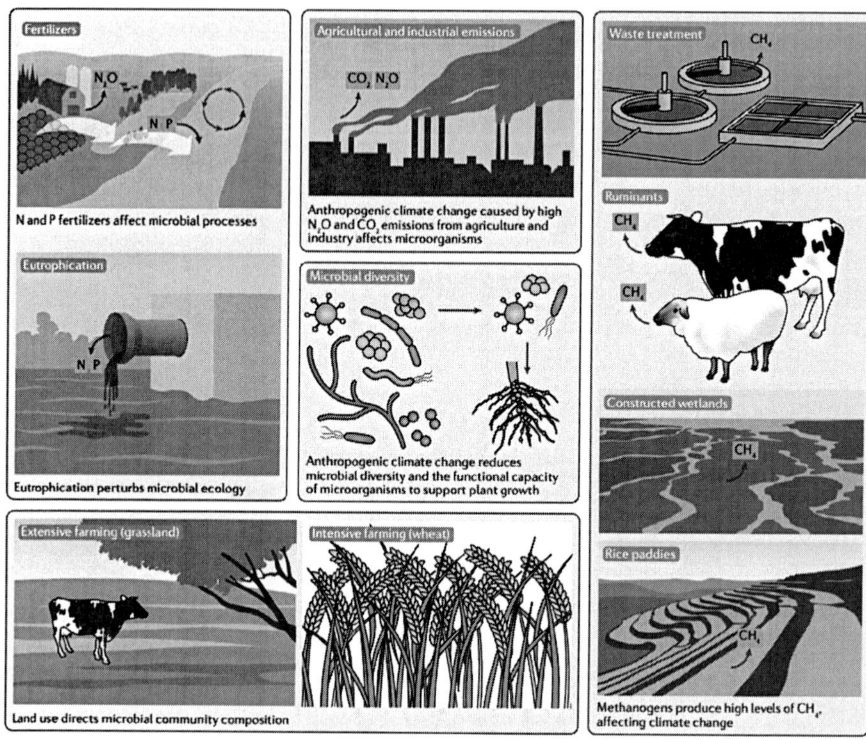

Fig. 1: Farming and various human practices that impact microorganisms

Communicable illnesses

Climate change influences the occurrence and propagation of diseases in both marine and terrestrial ecosystems, contingent upon various socioeconomic, environmental, and host-pathogen-specific elements. To comprehend disease spread and formulate effective control measures, a comprehensive understanding of pathogen ecology, vectors, hosts, and the effects of dispersal and environmental conditions is crucial. For instance, the rise in sea surface temperatures is closely associated with coral diseases, impacting aquaculture, livestock, and crop health, thereby posing a threat to global food production. Human activities, including population growth and transportation, coupled with climate change, contribute to heightened antibiotic resistance in pathogens and the dissemination of waterborne and vector-borne diseases, consequently elevating the risk of illnesses among humans, animals, and plants (Harvell *et al.*, 2002).

Climate change poses a particular risk to vector-borne, foodborne, airborne, waterborne, and other environmental pathogens. In the case of vector-borne diseases, climate change is expected to impact the distribution of vectors, influencing the geographical areas where diseases are transmitted and the effectiveness with which vectors transmit pathogens. The efficiency of transmission is contingent on the duration between a vector feeding on an infected host and the vector becoming infectious. Warmer temperatures can significantly reduce this duration, providing more opportunities for transmission within the vector's lifespan. Certain vector-borne diseases, such as bluetongue, a viral disease with significant economic implications for livestock, have already surfaced in Europe as a result of climate change. Predictions suggest larger and more frequent outbreaks in the future (Jones *et al.*, 2019).

Biodiversity loss

The health of an ecosystem relies on biodiversity, which is highly susceptible to climate change. Because biodiversity contributes numerous services within the ecosystem, even a slight reduction in the population of species can have adverse effects on the stability and maintenance of the ecosystem (Gaston and Fuller, 2008). Microbes play a major role in influencing global greenhouse gas dynamics, exhibiting rapid responses to climate shifts. Their impact on the environment can be either beneficial or detrimental. Consequently, understanding these unseen microorganisms is essential for comprehending climate change. Investigating how these microbes contribute to climate change, both in terms of producing and consuming greenhouse gases, as well as how they are influenced by climate shifts, becomes important for a comprehensive study (Cavicchioli *et al.*, 2019).

The preservation of microbial diversity is crucial for the maintenance of soil health and quality, as well as for the execution of the soil's role as a living system that upholds biological productivity and fosters the health of plants and animals. Regrettably, the diminution of microbial diversity is linked to a corresponding decline in functional capacity. The phenomenon of climate change constitutes a main catalyst for the decline in biodiversity across local to global scales, potentially inducing consequential modifications to ecosystem functioning and services (Wu *et al.*, 2022).

Ecology seeks to comprehend the processes responsible for the generation and preservation of biodiversity across temporal and spatial dimensions. Climate exerts a significant influence on the structure and interactions within communities, as well as on their functional attributes (Jain *et al.*, 2020). From the onset, it is imperative to recognize that the impacts of climate change on microbial communities within terrestrial and aquatic environments are intricate and occasionally paradoxical, influenced by both deterministic and stochastic factors. Notably, alterations in temperature have been extensively documented to exert effects on microbial biodiversity across multiple dimensions, including but not limited to geographical range, phenology, and distribution or abundance (Guo *et al.*, 2018). Microbial metabolic activity, population growth rates, and the number of species all show an exponential increase as temperature rises. Notably, the diversity of bacteria and fungi tends to decrease as global latitude increases, corresponding to a decline in temperatures (Zhou *et al.*, 2016). Temperature variations can influence microbial biodiversity through various mechanisms. One frequently observed mechanism is that elevated temperatures stimulate increased metabolic rates, leading to longer population doubling times. This, in turn, affects ecological and evolutionary processes, such as mutation, speciation, and various interactions like parasitism, predation, competition, or mutualism (Yuan *et al.*, 2021 and Guo *et al.*, 2019). Consequently, if temperature is a driving factor for microbial diversity, we can anticipate that ecosystems experiencing warming trends, ranging from tropical to alpine forests, will become more diverse and dynamic. This heightened diversity may enhance essential processes like decomposition, nutrient cycling, and carbon sequestration (Zhou *et al.*, 2016).

Climate change has the potential to impact various environmental aspects, including seawater acidification, hypoxia, CO_2 accumulation, alterations in salinity, and shifts in sea level (Lafferty, 2017). These alterations disrupt marine ecosystems, particularly the under-explored microbial biodiversity, which has not received thorough research attention for many years. Marine microorganisms constitute 90% of the biomass in the ocean. They serve as the foundational elements in marine food webs and play a crucial role in significant biogeochemical cycles (Hillebrand *et al.*, 2018). Despite their essential role, the majority of ocean observation programs and research studies often overlook microbes (Canonico *et al.*, 2019). The limited studies that do exist primarily highlight shifts in the microbial biodiversity of the ocean, primarily attributed to rising water temperatures and sea acidification (Bunse *et al.*, 2016 and Kling *et al.*, 2020).

Researchers have identified 25 viruses, 33 bacteria, 23 protists, and 21 metazoans as the primary agents responsible for causing diseases in marine organisms such as plants, corals, molluscs, crustaceans, echinoderms, fishes, turtles, and mammals (Lafferty, 2017). It's important to note that a significant portion of the marine microbial diversity is still undiscovered, with less than 0.1% currently described (Abirami *et al.*, 2021).

Altered ecosystems

Numerous studies over several decades have investigated the connection between microorganisms and their respective ecosystems. The introduction of sequencing and metagenomics has revealed a vast array of microorganisms existing in aquatic, terrestrial, and urban settings. Although researchers can identify the microbes in these environments, there remains limited knowledge regarding their activities in each setting and the potential effects of climate change on microbial communities and their ecosystem functions (American Society for Microbiology, 2022). Changes in temperature, humidity, or elemental flux caused by climate change can significantly influence the structure of microbial communities and their metabolic activity (Xue *et al.* 2016; Zhou *et al.* 2012; Woodcroft *et al.* 2018).

Aquatic ecosystem

Over 70% of Earth's surface is covered by aquatic environments, including marine, freshwater, and intertidal ecosystems. These habitats support a wide variety of animal, plant, and microbial species, playing a crucial role in human well-being and health. Aquatic systems serve as essential sources of drinking water, cleaning, and food. Similar to terrestrial environments, climate change significantly affects the temperature dynamics in aquatic ecosystems. The sea surface temperature worldwide has consistently risen, as reported by NOAA (National Oceanic and Atmospheric Administration, 2022). Elevated ocean temperatures may result in the development of more powerful storms and increased flooding, reduced marine life and biodiversity, and changes in the elemental flux caused by microorganisms in the ocean. Climate change modifies the carbon reservoirs in the ocean. The marine carbon pump employs microbial activities to store substantial carbon quantities in the deep sea, playing a crucial role in controlling atmospheric CO_2 concentrations. Warming has the potential to decrease the capacity of oceans to store carbon (Laufkötter *et al.* 2016). The cycling of carbon in the oceans is primarily influenced by the activities of the microbial loop and viral shunt. However, our knowledge about microorganisms in deep ocean systems is currently limited (Azam *et al.* 1983).

Terrestrial ecosystem

The dynamics of microbial community structure can shift due to the selection of rapidly proliferating bacteria, leading to changes in diversity (Guo *et al.*, 2018). Prolonged exposure of a grassland ecosystem to elevated levels of carbon dioxide (CO_2) significantly changed the composition, structure, and potential interactions of microbial communities in the soil of the grassland (He *et al.*, 2010, Zhou *et al.*, 2010 and Zhou *et al.*, 2011). For the microorganisms identified in soil, there is still much uncertainty regarding the precise functions carried out by each species within the microbial community. This knowledge gap restricts our capacity to anticipate the responses of these communities to climate change-induced shifts. Past research has fallen short in comprehending the changes in both genetic and functional diversity of microbial communities caused by warming and its interactions with other environmental factors. To uncover the activities of these microbes in their ecosystem and their mutual interactions, investigations that delve into causation, rather than solely depending on correlation, are imperative for understanding the mechanisms steering alterations in microbial communities (Nash Suding *et al.*, 2003).

Mitigation of climate change through microbial processes

Enhancing the comprehension of microbial interactions is crucial for developing strategies to mitigate and manage the impacts of climate change. For instance, insights into how mosquitoes respond to the bacterium Wolbachia, a common arthropod symbiont, have led to the reduction of Zika, dengue, and chikungunya virus transmission. This has been achieved by introducing *Wolbachia* into populations of *Aedes aegypti* mosquitoes and releasing them into the environment. In agriculture, advancements in understanding the ecophysiology of microorganisms that convert N_2O into harmless N_2 offer options for mitigating emissions. Strategies such as manipulating the microbiota in the rumen and breeding programs targeting host genetic factors to influence microbial community responses are being explored to reduce methane emissions from cattle. Biochar, derived from the thermochemical conversion of biomass under oxygen limitation, plays a role in improving the stabilization and accumulation of organic matter in iron-rich soils. Its impact includes enhanced organic matter retention by minimizing microbial mineralization and reducing the influence of root exudates in releasing organic material from minerals. This promotes grass growth, ultimately curbing carbon release (Weng *et al.*, 2017).

Conclusion

Global warming and climate change pose significant challenges worldwide, impacting various ecosystems on Earth, including plants, animals, and microbes. These changes influence the structure of microbial communities and their crucial metabolic activities, leading to alterations in the Earth's biogeochemical cycles. Microorganisms can accelerate global climate change by releasing greenhouse gases during the decomposition of organic matter. However, they can also mitigate climate change by converting greenhouse gas emissions into organic forms usable for their sustenance. Soil bacteria exhibit a modest response to global climate change and warming, whereas soil and root-associated fungi undergo more significant alterations. Elevated temperatures also impact fundamental ecological processes, geographical distribution, and the biodiversity, extinction, and degradation of aquatic species. Reductions in species populations have adverse effects on ecosystem sustainability.

References

Abirami, B., Radhakrishnan, M., Kumaran, S. and Wilson, A., 2021. Impacts of global warming on marine microbial communities. Science of The Total Environment, 791, p.147905.

American Society for Microbiology, 2022. Microbes and Climate Change-Science, People & Impacts: Report on an American Academy of Microbiology Virtual Colloquium held on Nov. 5, 2021.

Azam, F., Fenchel, T., Field, J.G., Gray, J.S., Meyer-Reil, L.A. and Thingstad, F.J.M.E.P.S., 1983. The ecological role of water-column microbes in the sea. Marine ecology progress series. Oldendorf, 10(3), pp.257-263.

Bunse, C., Lundin, D., Karlsson, C.M., Akram, N., Vila-Costa, M., Palovaara, J., Svensson, L., Holmfeldt, K., González, J.M., Calvo, E. and Pelejero, C., 2016. Response of marine bacterioplankton pH homeostasis gene expression to elevated CO2. Nature Climate Change, 6(5), pp.483-487.

Canonico, G., Buttigieg, P.L., Montes, E., Muller-Karger, F.E., Stepien, C., Wright, D., Benson, A., Helmuth, B., Costello, M., Sousa-Pinto, I. and Saeedi, H., 2019. Global observational needs and resources for marine biodiversity. Frontiers in Marine Science, 6, p.367.

Cavicchioli, R., Ripple, W. J., Timmis, K. N., Azam, F., Bakken, L. R., Baylis, M., et al. (2019). Scientists' warning to humanity: microorganisms and climate change. Nature Reviews Microbiology, 17(9), 569–586 Retrieved from https://doi.org/10.1038/s41579-019-0222-5. doi:10.1038/s41579- 019-0222-5.

Galfalk, M., Olofsson, G., Crill, P. & Bastviken, D. Making methane visible. Nat. Clim. Change 6, 426–430 (2016)

Gaston, K.J. and Fuller, R.A., 2008. Commonness, population depletion and conservation biology. Trends in ecology & evolution, 23(1), pp.14-19.

Guo, X., Feng, J., Shi, Z., Zhou, X., Yuan, M., Tao, X., Hale, L., Yuan, T., Wang, J., Qin, Y. and Zhou, A., 2018. Climate warming leads to divergent succession of grassland microbial communities. Nature Climate Change, 8(9), pp.813-818.

Guo, X., Zhou, X., Hale, L., Yuan, M., Ning, D., Feng, J., Shi, Z., Li, Z., Feng, B., Gao, Q. and Wu, L., 2019. Climate warming accelerates temporal scaling of grassland soil microbial biodiversity. Nature Ecology & Evolution, 3(4), pp.612-619.

He, Z., Xu, M., Deng, Y., Kang, S., Kellogg, L., Wu, L., Van Nostrand, J.D., Hobbie, S.E., Reich, P.B. and Zhou, J., 2010. Metagenomic analysis reveals a marked divergence in the structure of belowground microbial communities at elevated CO2. Ecology Letters, 13(5), pp.564-575.

Hillebrand, H., Brey, T., Gutt, J., Hagen, W., Metfies, K., Meyer, B. and Lewandowska, A., 2018. Climate change: warming impacts on marine biodiversity. Handbook on marine environment protection: science, impacts and sustainable management, pp.353-373.

Harvell, C. D. et al. Climate warming and disease risks for terrestrial and marine biota. Science 296, 2158–2162 (2002)

IPCC. Climate change 2022: impacts, adaptation, and vulnerability. Contri-bution of Working Group II to the Sixth Assessment Report of the Inter-governmental Panel on Climate Change, in press. Cambridge UniversityPress, Cambridge, United Kingdom.

Jain, P.K., Purkayastha, S.D., De Mandal, S., Passari, A.K. and Govindarajan, R.K., 2020. Effect of climate change on microbial diversity and its functional attributes. In Recent Advancements in Microbial Diversity (pp. 315-331). Academic Press.

Jones, A. E. et al. Bluetongue risk under future climates. Nat. Clim. Change 9, 153–157 (2019)

Kling, J.D., Lee, M.D., Fu, F., Phan, M.D., Wang, X., Qu, P. and Hutchins, D.A., 2020. Transient exposure to novel high temperatures reshapes coastal phytoplankton communities. The ISME journal, 14(2), pp.413-424.

Lafferty, K.D. Marine Infectious Disease Ecology. Annu. Rev. Ecol. Evol. Syst. 2017, 48, 473–496.

Laufkötter, C., Vogt, M., Gruber, N., Aumont, O., Bopp, L., Doney, S.C., Dunne, J.P., Hauck, J., John, J.G., Lima, I.D. and Seferian, R., 2016. Projected decreases in future marine export production: the role of the carbon flux through the upper ocean ecosystem. Biogeosciences, 13(13), pp.4023-4047.

Nash Suding, K., Goldberg, D.E. and Hartman, K.M., 2003. Relationships among species traits: separating levels of response and identifying linkages to abundance. Ecology, 84(1), pp.1-16.

Woodcroft, B.J., Singleton, C.M., Boyd, J.A., Evans, P.N., Emerson, J.B., Zayed, A.A., Hoelzle, R.D., Lamberton, T.O., McCalley, C.K., Hodgkins, S.B. and Wilson, R.M., 2018. Genome-centric view of carbon processing in thawing permafrost. Nature, 560(7716), pp.49-54.

Wu, L., Zhang, Y., Guo, X., Ning, D., Zhou, X., Feng, J., Yuan, M.M., Liu, S., Guo, J., Gao, Z. and Ma, J., 2022. Reduction of microbial diversity in grassland soil is driven by long-term climate warming. Nature microbiology, 7(7), pp.1054-1062.

Weng, Z. H. et al. Biochar built soil carbon over a decade by stabilizing rhizo deposits. Nat. Clim. Change 7, 371–376 (2017).

Xue, K., M. Yuan, M., J. Shi, Z., Qin, Y., Deng, Y.E., Cheng, L., Wu, L., He, Z., Van Nostrand, J.D., Bracho, R. and Natali, S., 2016. Tundra soil carbon is vulnerable to rapid microbial decomposition under climate warming. Nature Climate Change, 6(6), pp.595-600.

Yuan, M.M., Guo, X., Wu, L., Zhang, Y.A., Xiao, N., Ning, D., Shi, Z., Zhou, X., Wu, L., Yang, Y. and Tiedje, J.M., 2021. Climate warming enhances microbial network complexity and stability. Nature Climate Change, 11(4), pp.343-348.

Zhou, J., Deng, Y., Luo, F., He, Z., Tu, Q. and Zhi, X., 2010. Functional molecular ecological networks. MBio, 1(4), pp.10-1128.

Zhou, J., Deng, Y., Luo, F., He, Z. and Yang, Y., 2011. Phylogenetic molecular ecological network of soil microbial communities in response to elevated CO2. MBio, 2(4), pp.10-1128.

Zhou, J., Deng, Y.E., Shen, L., Wen, C., Yan, Q., Ning, D., Qin, Y., Xue, K., Wu, L., He, Z. and Voordeckers, J.W., 2016. Temperature mediates continental-scale diversity of microbes in forest soils. Nature communications, 7(1), p.12083.

Zhou, J., Xue, K., Xie, J., Deng, Y.E., Wu, L., Cheng, X., Fei, S., Deng, S., He, Z., Van Nostrand, J.D. and Luo, Y., 2012. Microbial mediation of carbon-cycle feedbacks to climate warming. Nature Climate Change, 2(2), pp.106-110.

10

Mitigation and Adaptation Strategies for Climate Change and Processing and Post-Harvest of Crops

Praveen Gidagiri [1*]***, Shameena S*** [1] ***and Gouthami Y*** [2]

[1]*Department of Postharvest Management, College of Agriculture, Vellayani Kerala Agricultural University, Kerala*

[2]*Department of Postharvest Management, College of Horticulture, Bagalkot University of Horticultural Sciences, Bagalkot, Karnataka*

Abstract

Climate change significantly impacts the pre-harvest physiology of various horticultural products, subsequently affecting their postharvest quality and storage potential. Changes in precipitation, temperature, carbon dioxide levels, solar radiation, ozone, and UV radiation directly influence horticultural production. Higher temperatures directly affect photosynthesis, altering sugar, organic acid, and flavonoid levels, as well as firmness and antioxidant activity. Elevated levels of atmospheric carbon dioxide can impact the quality of crops after harvest, resulting in issues such as tuber malformation, increased occurrence of common scab, and alterations in potato sugar content, among others. Elevated ozone levels can reduce photosynthesis efficiency, inhibiting growth and biomass accumulation. Climate change indirectly impacts food quality, affecting related markets and industries. Ensuring a well-managed cold chain supply is essential for mitigating metabolic changes and minimizing the growth of foodborne pathogens in harvested produce. Understanding climate change's impact on these aspects is vital for global food security. This chapter offers a thorough examination of how climate change affects postharvest physiology and the overall quality of fruits and vegetables. It underscores the importance of strategic navigation in addressing these implications.

Keywords: *Climate change, Postharvest quality, Elevated CO2, Temperature effects, Water stress*

Introduction

Climate change may advance more rapidly than anticipated, introducing novel environmental conditions for cultivating horticultural crops. The rise in atmospheric CO_2 levels from non-renewable energy sources combustion drives global average temperatures, altering weather patterns worldwide with regional variations (Stocker *et al.*, 2013). Currently, certain regions are already experiencing the effects of extreme weather events such as heat waves or frost, while prolonged droughts pose a growing threat to global food security elsewhere. Climate changes may yield both positive and negative outcomes, but they will undoubtedly alter production conditions and product quality (Kaufmann and Blanke, 2017).

Elevated temperatures can enhance the ability of air to hold water vapor, leading to increased water demand. This process can elevate evapotranspiration rates, potentially depleting soil water reserves and causing water stress in plants during dry periods. Fruit production is particularly vulnerable to water stress, especially in regions where trees lack irrigation. Numerous studies indicate that water stress diminishes crop yields and hastens fruit ripening (Henson, 2008).

Quality parameters of fruit and vegetables

The quality of fruits and vegetables varies depending on the species and the edible parts, such as fruit, leaves, or roots. These qualities can be categorized according to consumer perception, such as appearance (size, shape, color), texture (firmness, crispness, elasticity), flavor (sweetness, bitterness, astringency), and odor. Another set of quality traits relates to nutritional value, including minerals, vitamins, antioxidants, and other compounds from the plant's secondary metabolism. The synthesis and metabolism of specific compounds, through various metabolic pathways, regulate these products' perceived and nutritional quality.

Elevated temperatures, along with increased ozone and CO_2 concentrations, significantly impact the postharvest quality of fruits and vegetables. Higher atmospheric ozone levels enhance ascorbic acid content while reducing ester production in strawberries. Additionally, elevated CO_2 levels directly affect the incidence of common scab, tuber malformation, and changes in reduced sugar content in stored potatoes (Hogy and Fangmeier, 2009). Exposure of tomato fruits to ozone increases lutein, total carotenoids, and lycopene concentrations (Moretti *et al.* 2010). Hence, climate change-induced shifts in atmospheric temperature, CO_2, and ozone levels can alter the postharvest quality of various fresh fruits and vegetables.

Pre-harvest climatic factors affecting postharvest quality of produce

Environmental conditions during growth, such as extreme temperatures, wind, carbon dioxide levels, rainfall, and hail, play a crucial role in determining the yield and quality of agricultural produce during storage. The quality of the edible portion of a plant at the time of harvest largely influences its behavior throughout the postharvest phase. For fruits, factors like their position within the canopy and exposure to sunlight or shade can significantly impact the development of various bioactive compounds, thus influencing their quality during storage and distribution. Even fruits harvested simultaneously from the same tree display variations in physiological responses due to the complex interactions among environmental factors influencing the horticultural commodity.

Climate change can substantially affect the postharvest quality of fruits and vegetables. The postharvest period refers to the time between the harvest of crops and their consumption or processing. Changes in temperature, precipitation patterns, and extreme weather events associated with climate change can affect various aspects of postharvest quality. Below are several ways climate change might affect the postharvest quality of fruits and vegetables.

Temperature effects: Temperature significantly impacts on all physiological and biochemical processes involved in plant growth and yield. Elevated temperatures during field conditions can detrimentally affect the lifespan and quality of produce. High temperatures accelerate the consumption of stored carbohydrates in fruits, vegetables, and flowers due to heightened respiration rates. As a result, plants respire more rapidly, resulting in reduced storage or vase life.

For instance, tomatoes tend to ripen rapidly both on and off the plant when exposed to high temperatures. Oranges cultivated in tropical regions often exhibit higher sugar content and total soluble solids (TSS) compared to those grown in temperate regions. However, oranges from tropical climates typically have a green color and less efficient peeling, which can be attributed to the lower diurnal temperature variation experienced in these regions.

Increased Temperature: High-temperature-induced heat stress significantly increases membrane permeability of the affected fruits and vegetables (Moretti *et al.* 2010) and accelerates metabolic processes in fruits and vegetables, leading to faster rates of ripening and senescence. This may lead to a shorter shelf life and reduced quality. Increased temperature during development also substantially influences water relations, photosynthesis, membrane stability, certain hormones, and primary and secondary metabolites. Higher concentrations of certain polyamines concerning high-temperature stress during preharvest growth stages significantly influence edible produce's postharvest quality and storage potential (Malik and Singh 2006).

Temperature extremes: Extreme temperatures, whether high or low, can induce physiological disorders like sunburn or chilling injury, impacting the appearance and texture of fruits and vegetables.

Water availability: Water is the foremost limiting factor in agricultural crop production, with precipitation being its uncontrolled form that significantly impacts evolving weather patterns. Characterized by unpredictable climatic drivers, rain influences plant growth and development within specific regions. These drivers encompass fluctuations in temperature, humidity, flooding, and winds. Additionally, non-climatic factors linked to rain, such as pests and pathogens, can devastate standing crops or leave lasting effects on edible commodities.

Too much water, either through rain or irrigation, causes leafy vegetables like lettuce to become more brittle. Similarly, carrot roots split due to heavy irrigation after 90 days, followed by drilling. However, if heavy irrigation is preceded by minimal irrigation for the first 120 days, it prevents splitting and improves skin color with a slight decrease in yield (McGarry, 1993). In carrots, preharvest drought stress decreases membrane integrity in roots, which may increase dehydration during postharvest storage (Shibairo *et al.* 1998). The researchers found that fruits of the plants stressed for 45 days have higher peel redness, firmness, titratable acidity, and severity of internal pulp darkening. Water scarcity to fruit plants during growth may positively or negatively influence the taste and aroma of fruits during postharvest storage. Precipitation also positively or negatively affects the edible produce.

Changes in precipitation: Variations in precipitation patterns can affect water availability for irrigation, affecting the postharvest quality of fruits and vegetables. Inconsistent water supply may result in water stress, affecting the texture and flavor of the produce.

Drought stress: Prolonged drought conditions can lead to water stress in plants, reducing yields and affecting the overall quality of harvested crops.

Table 1: Effect of drought (deficit irrigation) on postharvest quality of produce

Commodity	Inferences	References
Blueberries	Regulated deficit irrigation (50% ETa) showed increased fruit firmness, SSC, and lower total acidity, while moderate irrigation (75% Eta) produced similar-quality fruit as in control (100% ETs)	Lobos *et al.*, 2016
Date palm	Deficit irrigation (70% ETc at 100 mm evaporation interval) did not affect the yield and quality of the produce but increased POD and PPO activities.	Alikhani-Koupaei *et al.*, 2018
Grapefruit	Deficit irrigation enhanced ripening index with increased SSC	Romero-Trigueros *et al.*, 2017
Pomegranate	Increased bioactive compounds, without affecting size and marketable yield with improved quality and health promoting attributes of fruits	Galindo *et al.*, 2017; Pena-Estevez *et al.*, 2016
Sugar apple	Increased antioxidants, ascorbic acid, and sugar contents	Kowitcharoen *et al.*, 2018
Tomato	Bioactive compounds were significantly influenced by regulated deficit irrigation, while changes in antioxidants were cultivar-dependent	Bogale *et al.*, 2016

CO_2 levels

Altered atmospheric CO_2 levels: Rising atmospheric carbon dioxide (CO_2) levels can influence the nutritional content of fruits and vegetables. Nutrient composition changes may impact the produce's overall quality and nutritional value. The elevated levels of CO_2, in combination with other abiotic stresses, negatively impact different crop plants (Kumar *et al.* 2019; Lamichaney *et al.* 2019). The elevated CO_2 may have positive effects on crops under simulated conditions.

Hogy and Fangmeier (2009) reported that high CO_2 concentration in the atmosphere (50% higher) resulted in a notable increase in tuber malformation (around 63%)and diminished processing quality.

Table 2: Effect of elevated CO_2 on postharvest quality of produce

Commodity	Inferences	References
Button mushroom	The browning index was reduced with maintained flavor and enhanced antioxidant enzyme activities under elevated CO_2 conditions.	Lin *et al.*, 2017
Oyster mushroom	Higher CO_2 (30%) increased shelf life with enhanced antioxidant enzyme activities and better sensory quality.	Li *et al.*, 2013; Zhang *et al.*, 2015

Commodity	Inferences	References
Persimmon	Soluble tannin contents were reduced in CO_2-treated fruits, and postharvest storage life was enhanced with increased antioxidant enzyme activities.	Min *et al.*, 2018
Strawberry	The storage life of fresh fruits is prolonged for 12 days. It also delayed the loss of chlorophyll and anthocyanin contents.	Li *et al.*, 2018; Li *et al.*, 2019; Shin *et al.*, 2008

Storage conditions

Energy demands: Fluctuations in temperature and humidity can influence the energy needed to store and transport fruits and vegetables. The heightened energy requirements for cooling and preservation may have implications for the cost and sustainability of postharvest handling.

Strategies for mitigating climate change impacts on the postharvest quality of horticultural produce

To mitigate the adverse impacts of climate change on postharvest produce quality, it's imperative to adopt effective strategies. Robust measures include integrating advanced postharvest technologies, adapting agricultural practices, and breeding resilient crop varieties capable of thriving in evolving climates. These strategies are aimed at bolstering the resilience and sustainability of the horticultural sector. Additionally, global, regional, and local collaborative efforts are essential to collectively tackle the complex challenges climate change brings on postharvest horticultural produce. Here, we delve into proactive postharvest strategies aimed at preserving and enhancing crop quality amidst ongoing climate shifts.

Adoption of improved Postharvest processing technologies

- **Drying and Dehydration:** Climate change may bring about changes in precipitation patterns, affecting the drying of crops. Improved drying technologies can help reduce moisture content, preventing the growth of mold and fungi during storage.

- **Canning and Freezing:** These technologies help preserve the nutritional quality of fruits and vegetables while extending their shelf life, providing a buffer against climate-induced challenges.

Cold chain management

- **Efficient transportation:** Maintaining a seamless cold chain during transportation from farm to market helps prevent temperature fluctuations that can accelerate spoilage.

- **Refrigerated transportation:** Using refrigerated trucks and containers helps to maintain the freshness and quality of perishable goods during transit.

Adoption of storage facilities

- **Climate-controlled storage:** Investing in climate-controlled storage facilities can help regulate temperature and humidity, mitigating the impact of temperature extremes on the postharvest quality of fruits and vegetables.
- **Modified atmosphere storage:** Implementing modified atmosphere storage systems can enhance the shelf life of produce by controlling O_2 and CO_2 levels, reducing respiratory activity, and slowing down the ripening process.

Conclusion

This chapter examines the multifaceted effects of climate change on the quality of horticultural produce after harvest. Factors like temperature variations, altered precipitation, and CO_2 levels significantly affect yield and nutritional composition. Elevated temperatures and water stress directly impact attributes like firmness and flavor, posing challenges in storage. Despite adversities, opportunities emerge, such as positive effects observed under elevated CO_2. By exploring proactive postharvest approaches like advanced processing techniques, efficient cold storage management, and the use of climate-controlled facilities, practical methods for preserving and safeguarding crop quality emerge. By embracing these strategies, the horticultural industry can bolster its resilience and sustainability, thereby contributing to global food security in the face of a changing climate.

References

Alikhani-Koupaei, M., Fatahi, R., Zamani, Z., and Salimi, S., 2018. Effects of deficit irrigation on some physiological traits, production and fruit quality of 'Mazafati' date palm and the fruit wilting and dropping disorder. Agric Water Manag. 209: 219–227.

Bogale, A., Nagle, M., Latif, S., Aguila, M., and Muller, J., 2016. Regulated deficit irrigation and partial root-zone drying irrigation impact bioactive compounds and antioxidant activity in two select tomato cultivars. Sci. Hortic. 213: 115–124.

Galindo, A., Calin-Sanchez, A., Grinan, I., Rodriguez, P., Cruz, Z. N., Giron, I. F., and Torrecillas, A., 2017. Water stress at the end of the pomegranate fruit ripening stage produces earlier harvest and improves fruit quality. Sci. Hortic. 226: 68–74.

Henson, R., 2008. The rough guide to climate change, 2nd edn. Penguin Books, London.

Hogy, P. and Fangmeier, A., 2009. Atmospheric CO_2 enrichment affects potatoes: 2. tuber quality traits. European J. Agronomy, 30: 85-94.

Kaufmann, H. and Blanke, M., 2017. Performance of three numerical models to assess winter chill for fruit trees- A case study using cherry as model crop in Germany. Reg. Environ. Chang. 17: 715–723.

Kowitcharoen, L., Wongs-Aree, C., Setha, S., Komkhuntod, R., Kondo, S., and Srilaong, V., 2018. Pre-harvest drought stress treatment improves antioxidant activity and sugar accumulation of sugar apple at harvest and during storage. Agric. Nat. Resour. 52(2): 146–154.

Kumar, A., Nayak, A. K., Das, B. S., Panigrahi, N., Dasgupta, P., Mohanty, S., and Pathak, H., 2019. Effects of water deficit stress on agronomic and physiological responses of rice and greenhouse gas emission from rice soil under elevated atmospheric CO_2. Sci. Total Environ. 650: 2032–2050.

Lamichaney, A., Swain, D. K., Biswal, P., Kumar, V., Singh, N. P., and Hazra, K. K., 2019. Elevated atmospheric carbon–dioxide affects seed vigour of rice (Oryza sativa L.). Environ. Exp. Bot. 157: 171–176.

Li, D., Li, L., Xiao, G., Limwachiranon, J., Xu, Y., Lu, H., and Luo, Z., 2018. Effects of elevated CO_2 on energy metabolism and γ-aminobutyric acid shunt pathway in postharvest strawberry fruit. Food Chem. 265: 281–289.

Li, D., Zhang, X., Li, L., Aghdam, M. S., Wei, X., Liu, J., and Luo, Z., 2019. Elevated CO_2 delayed the chlorophyll degradation and anthocyanin accumulation in postharvest strawberry fruit. Food Chem. 285:163–170.

Li, P., Zhang, X., Hu, H., Sun, Y., Wang, Y., and Zhao, Y., 2013. High carbon dioxide and low oxygen storage effects on reactive oxygen species metabolism in Pleurotus eryngii. Postharvest Biol. Technol. 85: 141–146.

Lin, Q., Lu, Y., Zhang, J., Liu, W., Guan, W., and Wang, Z., 2017. Effects of high CO_2 in-package treatment on flavor, quality and antioxidant activity of button mushroom (Agaricus bisporus) during postharvest storage. Postharvest Biol. Technol. 123: 112–11.

Lobos, T. E., Retamales, J. B., Ortega-Farias, S., Hanson, E. J., Lopez-Olivari, R., and Mora, M. L., 2016. Preharvest regulated deficit irrigation management effects on post-harvest quality and condition of V. corymbosum fruits cv. Brigitta. Sci. Hortic. 207: 152–159.

Malik, A. U. and Singh, Z., 2006. Improved fruit retention, yield and fruit quality in mango with exogenous application of polyamines. Sci. Hortic. 110: 167–174.

McGarry, A., 1993. Mechanical properties of carrots. Postharvest biology and handling of fruit, vegetables and flowers. Meeting of the Association of Applied Biologists, London.

Min, D., Dong, L., Shu, P., Cui, X., Zhang, X., and Li, F., 2018. The application of carbon dioxide and 1-methylcyclopropene to maintain fruit quality of 'Niuxin' persimmon during storage. Sci. Hortic. 229: 201–206.

Moretti, C. L., Mattos, L. M., Calbo, A. G., and Sargent, S. A., 2010. Climate changes and potential impacts on postharvest quality of fruit and vegetable crops: a review. Food Res. Int. 43(7): 1824–1832.

Pena-Estevez, M. E., Artes-Hernandez, F., Artes, F., Aguayo, E., Martinez-Hernandez, G. B., Galindo, A., and Gomez, P. A., 2016. Quality changes of pomegranate arils throughout shelf life affected by deficit irrigation and pre-processing storage. Food Chem. 209: 302–311.

Romero-Trigueros, C., Parra, M., Bayona, J. M., Nortes, P. A., Alarcon, J. J., and Nicolas, E., 2017. Effect of deficit irrigation and reclaimed water on yield and quality of grapefruits at harvest and postharvest. LWT-Food Sci. Technol. 85: 405–411.

Shibairo, S. I., Upadhyaya, M. K., and Toivonen, P. M. A., 1998. Influence of preharvest water stress on postharvest moisture loss of carrots (Daucus carota L.). J. Horti. Sci. Biotechnol. 73: 347–352.

Shin, Y., Ryu, J. A., Liu, R. H., Nock, J. F., Polar-Cabrera, K., and Watkins, C. B., 2008. Fruit quality, antioxidant contents and activity, and antiproliferative activity of strawberry fruit stored in elevated CO_2 atmospheres. J. Food Sci. 73(6): 339–344.

Stocker, T. F., Qin, D., Plattner, G. K., Tignor, M., Allen, S. K., Boschung, J., Midgley, P. M., Nauels, A., Xia, Y., and Bex, V., 2013. Climate Change 2013: The Physical Science Basis; Cambridge University Press: Cambridge, UK.

Zhang, L., Gao, J., Hu, H., and Li, P., 2015. The activity and molecular characterization of a serine proteinase in Pleurotus eryngii during high carbon dioxide and low oxygen storage. Postharvest Biol. Technol. 105: 1–7.

11

Mitigation and Adaptation Strategies for Climate Smart Extension

Manju Prem S.[1], Mohanraj M., Swadhin Priyadarsinee[3] and Sampriti Guha[4]

[1]Department of Agricultural Extension Education, College of Agriculture Vellayani, Kerala Agricultural University, Kerala
[2]Department of Agricultural Extension, College of Agriculture, Bangalore University of Agricultural Sciences, Bangalore, Karnataka
[3&4]Department of Agricultural Extension, Bidhan Chandra Krishi Viswavidyalaya, Mohanpur, West Bengal

---

Abstract

This book chapter explores how climate change significantly impacts agriculture, affecting crops, livestock, and overall farming practices. It highlights the challenges faced by vulnerable populations, especially those relying on isolated agricultural systems, due to altered weather patterns and extreme events. The chapter emphasizes the crucial role of agricultural extension services, which help farmers adapt to these challenges.

In addressing climate change, the chapter discusses practical strategies for agricultural extension officers. These include promoting climate-smart farming practices, providing weather forecasts and early warnings, supporting resilient livestock farming, and facilitating training workshops. The goal is to empower farmers with the knowledge and skills needed to adapt to changing climate conditions and promote sustainable agriculture. Collaboration between extension officers, policymakers, and farmers is emphasized for effective management of climate-related challenges. The chapter also stresses the importance of integrating climate information into agricultural extension services to assist farmers in making informed decisions. Furthermore, capacity building among extension agents is highlighted as a key factor in successfully implementing climate-smart agriculture practices and ensuring sustainable development in the face of climate change.

Keywords: *Climate-smart farming, Agricultural extension, Climate change adaptation, Vulnerability assessment, Sustainable agriculture and Capacity building.*

Introduction

Agriculture is the backbone of every country. It largely depends on climate and its changes. Over the last few years climate change has affected crops, cropping systems, patterns and the ecosystem hugely. Climate change is caused by greenhouse gases in the atmosphere, which results in global warming (Aydinalp and Cresser, 2008). The rising temperature, melting of glaciers, loss in vegetation cover, frequent occurrences of drought and floods are some of the issues that have raised concern among the environmentalists. The change in temperature and precipitation patterns alters the traditional growing seasons. Climate change is expected to influence crop and livestock production, hydrologic balances, input supplies and other components of agricultural systems (Adams *et. al.,* 1998). Vulnerability of climate change depends on physical, biological and socio-economic characteristics. Low-income populations dependent on isolated agricultural systems are particularly vulnerable to hunger and severe hardship. These populations are already barely food-sufficient, even the slightest decline in yields could be very harmful in these areas. The most negative effects are foreseen in dry land areas at lower latitudes and in arid and semi-arid areas, especially for those reliant on rain fed, non-irrigated agriculture (Aydinalp and Cresser, 2008). The rise in temperature cause stress in crops and reduce yields and productivity. Crops that are sensitive to temperature during the growth stages are affected a lot. A shift in growing season is being seen due change in temperature. Irregular and altered patterns of rainfall are severely affecting the crops and water supplies for livestock, which in turn has increased the occurrence of floods and drought, causing soil erosion, disrupting harvesting schedules. Occurrence of new pests and diseases are seen in the places having warmer temperature. Altered temperature has altered the life cycle of pests and intensity of infestations. Rising of sea water level and increased evaporation leads to intrusion of salt water into fresh water making irrigation water more saline harmful to crops. Livestock are sensitive to heat stresses which in turn reduce livestock productivity and increase mortality rates. The availability and quality of forage crops is also affected. Climate change has reduced the yield of major crops like wheat, rice and maize which has affected the global food supply. Due to alteration in the export and import of food grains there is a rise in the prices of food grains both at producer and consumer end.

The term "agricultural technology transfer" is used to describe the process of formally transferring new agricultural discoveries, improved practices or innovations that may result from research institutions into the agricultural sector. Agricultural extension is the process of carrying the technology of scientific agriculture to the farmers in order to enable the farmers to utilize the knowledge and a better economy (Altab *et. al.*, 2015). Agricultural extension has multiple goals, including transferring knowledge from global, national, and local researchers to farmers, helping them clarify their own goals and assessing their opportunities, educating them about decision-making processes, and promoting desirable agricultural development (Msuya *et. al.,* 2017).

Agricultural extension can play a role by ensuring that efforts towards increasing productivity are sustainable. Sustainable agricultural production ensures that current production activities do not compromise the production chances in the future. One way of achieving this is through agro-ecological practices (FAO, 2018). Agro-ecological practices are premised on empowering farmers as key agents of change while fostering co-creation of knowledge, integrating traditional, practical and local practices and skills for sustainable production (FAO, 2018).

Agricultural Extension Services play a crucial role in disseminating knowledge, technology and best practices to farmers. Adoption of climate resilient agricultural practices, enhance farmers, adaptive capacities and promote sustainable agriculture.

How agricultural extension contributes addressing climate change?

1. ***Dissemination of Climate-Smart Farming Practices***: Extension officers can provide information on climate-resilient crop varieties that are better adapted to changing climate conditions, such as drought-resistant or heat-tolerant crops. They can educate farmer about efficient water management techniques, including rainwater harvesting, drip irrigation, and soil moisture conservation to cope with changing precipitation patterns. Soil conservation practices like cover cropping, mulching and agroforestry can enhance soil fertility, prevent soil erosion and improve resilience to climate related challenges.

2. ***Weather Forecasting and Early Warning Systems:*** The extensive services provide farmers with up-to-date weather forecasts and climate information, allowing them to make informed decisions about planting, irrigation and pest management role. These services are aware of severe weather events such as floods or droughts, and set up early warning systems

3. ***Promoting Climate-Resilient Livestock Farming:*** The extension program provides guidance on livestock practices, including improved shelter, water supply management, and disease controls tailored to

changing climate. Educating livestock farmers on resistant food crops weather conditions and pasture rotations to help ensure constant forage for cattle even in drought conditions.

4. ***Capacity Building and Training:*** Extension officers can organize training workshops, demonstrations and field visits to educate farmers on climate smart methods and technologies. Facilitating farmer-to-farmer knowledge exchange and community learning can help spread successful climate adaptation practices at both local and regional levels.
5. ***Supporting Sustainable Agricultural Practices*:** Agro ecological practices, such as integrated pest management, integrated nutrient management, organic farming and biodiversity conservation, can empower biodiversity more resilient and reduce the need for chemicals. Extension workers can educate farmers on carbon farming practices, such as agroforestry and conservation agriculture. It can help remove carbon dioxide from the atmosphere and have reduced the effects of climate change.
6. ***Policy Advocacy and Research:*** Extension agencies can advocate for policies that encourage climate-smart agriculture, including promoting sustainable practices and integrating climate adaptation into agricultural policies and programs. Collaboration with research institutions, extension services can facilitate the dissemination of climate-related research findings to farmers, helping them adopt innovative and evidence-based practices.

Farmers can be provided with the necessary knowledge and skills, agricultural extension services empower them to adapt to climate change, build resilience, and contribute to sustainable agricultural development. Collaboration between extension officers, policymakers and farmers is essential in effective management of challenges posed by climate change.

The long term changes in temperature, precipitation and other atmospheric conditions on earth have caused climate change. Intervention of human activities, particularly emission of greenhouse gases, has significantly contributed to the warming of the planet. Agriculture has been profoundly impacted by climate change.

Climate change and its effects on agriculture

1. ***Temperature Changes*:** The sudden rise in temperature has affected the growth cycles of many crops, due to which there has been a shift in the time of sowing, transplanting and in fact harvesting schedules. Due

to increase in heat stress, there is decreased productivity in livestock population. Marine population migrate from hot areas to cold areas.

2. ***Altered Precipitation Patterns***: Climate change has affected the monsoon period. Late arrival of the monsoon period has disrupted the agricultural activities. The changes in rainfall intensity and frequency cause water stress among the crops. Due to low intensity rainfall, the ground water table has depleted. Frequent occurrences of drought and floods have disrupted the irrigation pattern of the crops.

3. ***Pests and Diseases:*** Proliferation of new pests and diseases are now seen in due to changes in temperature. Warmer temperatures allow pests and diseases to thrive in new regions where they were previously uncommon, leading to increased crop damage. Altered climate patterns can disrupt the life cycles of pests, affecting the timing and intensity of infestations.

4. ***Soil Health:*** Soil erosion has become more common due to changes in precipitation patterns and more intense storms. Erosion of top soil affects the nutritional status of both soil and plants. Decrease in nutritional content of soil affects the crop health and yields.

5. ***Water Availability:*** The amount of available water for irrigation is decreasing. A decrease in rainfall and melting glaciers reduce the water supply which is a critical factor for many crops. Rising sea levels leads mixing of saltwater into freshwater sources making freshwater sources, making irrigation water harmful for crops. Freshwater marine life also gets affected due to mixing of saline and marine water.

6. ***Shifts in crop suitability:*** Crops grown in tropical regions can also be grown in subtropical regions due to warmer climatic conditions. But it will be difficult to grow crops that belong to subtropical and polar regions due to change in temperature and climatic conditions. Farmers growing crops like coffee, tea, walnuts, cashew, and cocoa will face challenges.

7. ***Change in forage quality:*** Weather patterns such as drought, excessive rainfall can affect forage quality. Drought can lead to stressed plants with reduced nutritional content and excessive rainfall leach out nutrients from soil, affecting forage quality which in turn affects livestock health and production.

8. ***Socio-economic implications:*** Changes in temperature and precipitation patterns can lead to reduced agricultural productivity, affecting food

availability and increasing food prices. Small-scale farmers, particularly in developing countries, are vulnerable to climate-related crop failures, impacting their livelihoods and income. They have limited access to resources, health care and adapt to changing climatic conditions.

Loss in Bio-diversity

On land, species migrate into higher latitudes, as extreme weather events (such as floods, storms, hurricanes, and wildfires), and sea level rise invade the habitats of others. Species that cannot migrate or unable adapt to the new conditions face extinction. According to estimation 4% of terrestrial and 13% of marine species are at high risk of extinction in biodiversity hotspots within a warming scenario between 1.5 and 2 degrees Celsius. The shares of terrestrial and marine species extinction will rise 20 and 32% respectively if temperature increases above 3 degree Celsius. In marine ecosystems, there is increase in the average ocean temperature due to climate change. In addition, the absorption of carbon dioxide from the atmosphere increases the water's acidity. Marine ecosystems are extremely sensitive to even seemingly small variations in water conditions, with drastic outcomes expected. It is forecast that in a warming scenario of two degrees Celsius above pre-industrial levels, 99 per cent of coral reefs worldwide will vanish. As of 2022, the global temperature had increased by 1.2 degrees Celsius above the pre-industrial average.

Effects of drought

Droughts and desertification have a considerable impact on both people and the planet. A loss of soil productivity, crop yield volatility, poverty, famine, social conflicts, and mass migration are some of the issues that have resulted in drought. An economic damage 54 billion U.S. dollars is being accounted worldwide in 2020. More than 100 million people were impacted by droughts that same year. The frequency of agricultural and ecological droughts is going to increase over the coming decades as a change in global temperature is seen. With every rise in temperature the frequency of increase in drought will be seen. With every rise in 2 degree Celsius, 2 droughts every 10 years will be seen.

Adaptation Strategies

In light of changing weather patterns and harsh environmental conditions, various adaptation tactics in climate-changing agriculture are gaining importance. To reduce the effects of climate change on food output and sustainability, farmers round the globe must employ creative strategies. These techniques include- crop diversification, water management, use of resilient crop varieties, soil health practices, agroforestry, and access to climate information services (Challinor *et al*., 2014) (Rejesus *et al.,* 2011). To ascertain food security and agricultural resilience in a protean climate landscape such proactive strategies stands to be vital.

A tabular representation of different adaption techniques that farmers can implement to deal with climate change is given below:

Adaptation Strategy	**Description**
Crop Diversification	A variety of crops are planted to reduce the susceptibility to certain weather conditions and pests attacks.
Improved Water Management	Using water conservation techniques like drip irrigation, rainwater harvesting, and effective water storage systems.
Climate-Resilient Crop Varieties	Selecting or breeding crops having more resilience to adverse weather conditions, drought, and flooding.
Soil Health Management	Practices for improving soil structure and water retention capacity like no-till farming, cover crops, and the addition of organic matter should be used.
Agroforestry	To increase biodiversity, reduce soil erosion, and provide shade and windbreaks; trees and bushes must be incorporated with crops.
Climate Information Services	Use advisory services, climate data, and weather forecasts to make wise choices about when to sow and harvest the crops.

These strategies can support farmers to adapt to the difficulties caused by changing climatic conditions and foster sustainable agriculture. It's essential for farmers to amalgamate multiple strategies based on their specified geographical location and needs.

Mitigation strategies

The climate change mitigation methods in agriculture are of critical importance for addressing the dual problems of assuring food security and decreasing greenhouse gas emissions. Considering the extent to which agricultural industry contributes to global emissions, adopting sustainable practices has become crucial. These strategies cover a wide range of tactics, such as lowered livestock emissions and improved resource management (Smith *et al.,* 2008), carbon sequestration in soils through afforestation and organic farming (Lal, 2010), crop and livestock resilience to changing conditions (Lobell *et al.*, 2011), and energy-efficient farming techniques (Rezaei *et al.,* 2015). By focusing on these techniques, agriculture may contribute significantly to decrease climate change while supplying to the increasing global food demand.

Strategies for lowering climate change in agriculture are summarised in a tabular form:

Mitigation Strategy	Description
Reduced Greenhouse Gas Emissions	Enforcing practices to lower the emissions of methane (CH4) and nitrous oxide (N2O), like, improved manure management, optimized fertilizer use, and controlled irrigation.
Carbon Sequestration in Soils	Agroforestry, organic farming, reforestation, and other approaches can help sequester carbon in soil, recovering soil health and reducing climate change.
Sustainable Intensification	Enhancing resilience, reducing waste, and optimizing resource utilization to maintain a balance between increaded agricultural productivity and decreased environmental impact.
Crop and Livestock Resilience	Developing climate-resilient crop varieties and livestock breeds that can resist changing conditions, reducing vulnerability.
Energy-Efficient Farming Practices	Using energy-efficient technologies, renewable energy sources, and boosting farm operations to reduce greenhouse gas emissions.

Role of agricultural extension in climate adaptation

By facilitating farmers' access to the necessary knowledge and tools for implementing climate-smart agricultural practices, agricultural extension programs improve resilience and sustainability.

Extension Program Component	Description
Training Workshops	Conducting workshops to acquaint farmers with climate-smart agricultural methods and technologies.
Demonstration Farms	Setting up model farms to display climate-resilient approaches and provide farmers a chance to see and gain first-hand knowledge.
Information Materials	Delivering leaflets, flyers, and other materials laying out various adaptation measures regarding climate change amongst the farmers.
Mobile Apps and Web Portals	Designing user-friendly apps and websites to extract weather forecasts, guidance, and information about the climate in a simple and effective manner.
Farmer Field Schools	Developing farmer field schools for disseminating practical knowledge and information among farmers on climate-smart farming approaches.

These components of agricultural extension help farmers to obtain the necessary information they need to adopt climate-smart farming practices, boosting resilience and sustainability.

Incorporating climate information

Blending climate information and forecasts into agricultural extension services and decision-making processes is of utmost importance for several reasons. Scientific developments in weather forecasting has granted farmers some essential tools to improve resource management (Hansen *et al.*, 2011), reduce their impact on the environment (Lobell and Gourdji, 2012), and accustom to changing weather patterns. The importance of inculcating climatic data and forecasts in agricultural extension services and decision-making procedures is given below:

Importance	**Description**
Enhanced Resilience	The vulnerability towards droughts, floods, and storms will reduce when farmers can anticipate and prepare themselves for extreme weather conditions by using the climate information.
Optimized Resource Management	By strategically planning the time of irrigation, harvesting, and planting with the aid of climatic projections, farmers are able to better manage their resource allocation.
Improved Crop Selection	Seasonal weather forecasts guide farmers to select crop varieties which are compatible with the predicted weather, leading to more steady harvests.
Reduction of Environmental Impact	By reducing emissions, resource use, and carbon footprint, climate information can help support agriculture in having a less detrimental effect on the environment.
Access to Climate-Smart Practices	Farmers can gain knowledge about climate-smart farming approaches which tends to improve resilience and flexibility via extension services.
Risk Reduction	Crop failure, pest and disease outbreaks, and vulnerable weather conditions, all prove to be hazardous. The can be reduced by making wise decisions based on climate data.
Food Security	Assimilation of climatic data provides stable agricultural production, raising local and global food security.

Capacity building

Building capacity among extension agents and educators is critical for effectively addressing climate change challenges. The complexity and evolving nature of climate-related challenges necessitates a thorough understanding of scientific concepts, innovative solutions, and the ability to explain and implement strategies at the grassroots level, all of which necessitate a profound understanding of scientific principles. Extension people, such as agricultural extension officers, community workers, and other development practitioners, play an important role in sharing knowledge and encouraging communities to adopt climate-smart practices to overcome the challenges of climate change.

Importance of capacity building in climate smart agriculture for climate change

Capacity building is essential because it improves knowledge and skills to keep pace with the rapidly changing technological developments and also helps improve the feedback mechanisms from the field to the extension staff and researchers (Dwarakinath, 2006). Capacity building is needed as it helps in not just farmer training and strengthening the innovation process but also in building linkages between farmers and the various stakeholders involved in helping farmers. It is widely reported that public extension services frequently have some flaws, such as a lack of timely information and input supply, less accountability of public extension personnel, the blanket nature of recommendations, and the absence of extension personnel during office hours, which makes them less committed to the service.

Extension capacity building is sometimes disregarded in the drive to get the results of research and development goods to farmers. Building the capacity of agricultural extension personnel is critical to the success of climate smart agriculture efforts. Farmers in many developing nations, including India, face resource constraints and are underdeveloped. Agriculture and associated departments are already overburdened with their routine activities and do not place much emphasis on the increasing threat of climate change. As a result, increasing agricultural extension capability is critical for climate smart agriculture, poverty reduction, and environmental conservation. Implementing climate smart agriculture projects alone will not result in the necessary degree of agricultural growth because farmers, groups, and organizations must have the ability and responsibility to solve problems and improve their communities.

Experiential learning is an excellent way to develop good insights into the ways groups work and can be a very powerful method and appropriate in extension for climate smart agriculture particularly who work with groups. Cross visits, study tours, and Farmer Field School are effective methods of transferring information and technology to staff and farmers, particularly in remote areas. Farmers and staff can benefit from farmer field schools (FFS) by gaining knowledge, skills, good relationships, facilitator skills, communication skills, and experiences. Cross visits, study tours, and Farmer Field School are effective methods of transferring information and technology to staff and farmers, particularly in remote areas. Farmers and staff can benefit from Farmer Field Schools (FFS) by gaining knowledge, skills, good relationships, facilitator skills, communication skills, and experiences.

In order to build capacity in extension, mentoring is a crucial tool that should be carefully applied to climate-smart agriculture. Senior researchers and extension personnel with a wealth of experience serve as mentors. Those with greater expertise in Indigenous Technical Knowledge (ITK) and extension methodology are known as mentors. Mentoring is the process by which more seasoned employees impart knowledge, abilities, and attitudes to less seasoned extension workers. Scaling out impacts requires staff to develop their technical and extension skills, and one way to do this is by using experienced individuals as mentors, according to Millar and Connell (2005). They can offer the assistance that learners require to grow into responsible adults while they pick up new skills and adjust to changing circumstances. Strong communication, listening, analysis, feedback-giving, and negotiation skills are essential for mentors working with less seasoned individuals.

Requirement for capacity constructing in climate smart agriculture

In agriculture and related fields, capacity-building mostly deals with the creation or bolstering of official (government) and informal (NGOs, farmer groups, etc.) institutions, the private sector, and individuals. The intention is to better equip them to handle their policy and decision-making responsibilities and to carry out agricultural development programmes more effectively. This suggests decentralization all the way down to the local level as well as offering rewards for participation in and local community initiatives. To be able to train farmers, cooperatives, and other rural organizations to help with the consolidation of grassroots organizations, voluntary organizations and those representing the interests of the various farmer interest groups must be involved. Incorporating participatory techniques into the in-service training of agriculture and related department extension functionaries is a crucial addition to engaging local communities.

In general, agricultural and allied departments lack the full range of in-house expertise required to respond to the changes required for sustainable agriculture and climate smart agriculture. Modules to improve knowledge and skills in climate smart technologies should be included in ongoing capacity building and training programmes for extension functionaries. Climate smart agriculture involves several sectors and actors, and in order to plan for climate smart agriculture, the capacities of actors involved in different sectors must be strengthened, particularly at the local level, where actions are required. Agriculture officers and extension field functionaries are critical links in the translation of research findings into on-the-ground application. Field extension workers are essential to the development of agriculture at the field level. To effectively communicate with farmers, these individuals must have a thorough understanding of the issues, requirements, and potential solutions for promoting climate smart agriculture. Therefore, now more than ever, extension field workers and officials from the agriculture department will be the driving forces behind climate-smart agriculture. Localized changes can significantly improve the mitigation of the unfavorable effects of climate change. Building these employees' capacities on a subject that affects the entire world is essential, and every little effort counts.

Establishment of emergency management units in extension agencies

- Dissemination of innovations on best adaptation practice
- Training and re-training of extension staff to acquire new capacity in climate change management
- Improving feedback on climate change issues to the government and other interested parties

Essentially, the knowledge base of extension functionaries on climate change should be expanded so that they can train farmers. Despite the existence of several proven adaptation strategies, little to no effort has been made to develop appropriate training curriculum that incorporates the various adaptation strategies. As a result, there is a need to strengthen the capacity of extension functionaries on climate change issues such as causes, effects, and adaptation strategies.

Case Studies and best practices

Participatory Integrated Climate Services for Agriculture (PICSA)

Through a set of participatory decision-making tools, PICSA combines the analysis and communication of locally specific climate information with the identification and exploration of household response options. It is intended to assist farmers in making informed decisions improve their ability to manage and adapt to climate change (Dorward. *et al,* 2015).

PICSA has been effective in supporting farmers' decision-making in different countries and contexts. The question is why? This section draws from experience, reflections on and evaluations of multiple implementations and evaluations of the approach to discuss how it may have stimulated innovation in smallholder farming systems. We then consider future opportunities and challenges for the further use and development of the approach.

To provide evidence that PICSA promotes positive innovation and change in smallholder farming systems in various contexts and locations, and at scale. The emphasis on equipping farmers is a key reason for this success with a process, information, and tools to help them make informed decisions for their context. Importantly, the strategy does not seek to provide advice appropriate for all farmer contexts, which is an impossible task. Instead, it enables farmers to think about and plan for the future. Complexity of their own personal contexts and focuses on assisting instead of 'advising' the farmers. Scaling up PICSA is not always 'easy,' and it necessitates planning for different locations and farming systems, as well as close collaboration with key institutions to encourage ownership and build the capacity needed to support the process. There are significant opportunities and challenges in scaling the approach, ensuring sustainability, and maintaining the quality required to ensure farmers are empowered to make decisions that improve their livelihoods in the face of climate variability and change.

Using ICTs for micro-level drought preparedness in India

Due to its reliance on rainfall and surface water, India, particularly its agricultural sector, is extremely vulnerable to drought. To meet new challenges, the International Crops Research Institute for the Semi-Arid Tropics (ICRISAT) in India and its partners have experimented with various ICT tools such as video-conferencing, audio-conferencing, and mobile telephony for delivering agro-advisories related to weather and best agricultural practises. ICRISAT, in particular, recognises the importance of a drought vulnerability map with up-to-date weather information for planning micro-level climate adaptation strategies. These maps are color-coded, with red, orange, green, and yellow villages indicating their level of drought vulnerability. This allows potentially vulnerable farmers to assess their own vulnerability and make more informed decisions, providing a more solid foundation for designing adaptation strategies. During field surveys, the maps were distributed to rural communities, and farming communities were educated about drought preparedness using ICT tools such as mobile phones and video and audio conference systems. GIS maps and ICT tools are also used to provide agro-advisory services to farmers and to solve real-time field problems (Anon., 2007).

Public acceptance of renewable energy in Kenya

The importance of technology identification, prioritisation, and planning is demonstrated in the case study of public acceptance of renewable energy in Kenya. This case study was part of the TRANS risk project, which examined the significance of stakeholder engagement in Kenya's energy technology decision-making process. A Stakeholder engagement took the form of two consultations on the country's energy technology options. The information-choice questionnaire, which had previously been used to gauge public opinion on carbon dioxide capture and storage in the Netherlands, was used in the consultations. The questionnaire provided background information about the proposed technologies, avoiding the problem of 'pseudo opinions' being provided due to a possible lack of knowledge. In Kenya, the TRANS risk project included 100 interviews and took into account three technologies (wind, solar and geothermal) (Zwaan, *et al.*, 2018). According to the analysis of the interviews, the stakeholders provided the following important considerations about the proposed energy technologies:

(a) The positive use of land for both solar energy and agriculture.

(b) The increase in employment as a result of wind turbine construction.

(c) The financial benefits to local communities once wind turbines can be built in local communities.

(d) The low emissions of wind turbines.

(e) The need for infrastructure development, such as access roads enabling the maintenance and operation of wind turbines and

(f) The lower price of wind energy driving.

These consultations provided an important picture of public opinion. The advantages and disadvantages of each technology option influenced the government's strategy for overcoming obstacles and accelerating deployment. Public acceptance of renewable energy is critical for countries to meet their national energy demand and Paris Agreement targets. (Best-Waldhober, *et al.*, 2012).

Challenges and barriers

Agriculture is a complex and dynamic socio-ecological system that is subject to policy, economic, and climatic uncertainty. Site-specific Climate-Smart Agriculture (CSA) practises will be critical in increasing the income of the most vulnerable populations, mitigating the impact of climate change on food security, and strengthening smallholder resilience. Adoption of technologies such as CSA is likely to be aided by publicly available agricultural extension services.

These services provide vital information to farmers. However, several barriers to the adoption of climate-smart practises exist, including financial constraints, labour shortages, land and water scarcity, insufficient transportation resources, and low farmer organisation membership. Adoption barriers also include institutional, socioeconomic, and cultural factors. These include a lack of financial capital, a lack of knowledge, insufficient land tenure, market failures, and inadequate infrastructure.

Other factors to consider include the suitability of innovations for end users, labour requirements, access to external inputs, and the use of crop residues for animal feeding. Another significant impediment is the adoption process itself, particularly in terms of its social dimensions such as age, gender, and diversity.

Furthermore, the level of institutional support in a region can influence the rate at which CSA practises are adopted, particularly those requiring higher initial start-up costs or technical expertise. Infrastructure, extension services, and healthcare investments in agricultural communities can have a significant impact on farmers' willingness to take risks and, as a result, their adoption of new practises.

Future directions

The framework for Future-Climate, Current-Policy (FCCP) approach drew on all of the threads, but they aimed to simplify them enough to engage key stakeholders in a quick and understandable process. The goal has been to create an approach that is: sufficiently grounded and detailed to generate realistic draught plans; focused on the development of a set of achievable but challenging short-term actions; and geared towards addressing long-term change. The resulting 'Future-Climate, Current-Policy (FCCP) Framework' allows for a flexible debate that anchors proposed actions to an analysis of their likelihood of resulting in outcomes that mitigate future climate change impacts. There is explicit acknowledgement of the need to establish a link between the generation of new climate science and policy change on the ground. A series of meetings brought together leading climate scientists and sectoral policymakers in the fields of water management, urban infrastructure, agriculture and rural livelihoods, tea production, transportation, and fisheries. Policymakers and development practitioners in the region understand the "idea of climate change," but they are frequently unable to connect it to immediate policy options. This appears to be because climate change longer time horizons outstrip regular planning cycles (Evan, *et al.* 2020).

Two innovations attempt to bridge the gap between medium-term (40-year) change science and short-term planning. To begin, Climate Risk Narratives (CRNs) are used to distil the complex multidimensional nature of climate model projection data into three plausible quasi-quantitative climate scenarios. They are based on a synthesis of the most recent climate information, with a particular emphasis on communicating critical messages to audiences with varying levels of climate science capacity. More sophisticated and well-established scenario planning approaches, such as the two-two scenario method, are well adapted to later stages of the planning process, but are notable for their complexity.

Naturally, the frameworks are insufficient to support detailed sectoral planning, but they have the potential to serve as an 'anchoring' point for more dynamic adaptive pathway planning exercises. The complex reality of planning within challenging political environments, and with multiple dynamic processes interacting, means that no single 'act of planning' is sufficient; instead, an interlocked dynamic planning process is required.

A Scenario development, quantitative modelling, and scenario-guided policy and programme design play an important role in exploring options to address socioeconomic and environmental challenges across many sectors,' according to the report.

The FCCP Framework shows promise as a useful starting point for such detailed sectoral planning, and, more importantly, it may alleviate some of the policy paralysis that occurs when sector policymakers are confronted with climate change information. The real challenge, as many others have discovered, is finding an institutional 'home' that can serve as a long-term host for such innovative, responsive long-term planning.

Conclusion

Climate change is posing difficulties, for our farmers as the weather becomes increasingly unpredictable and disrupts farming practices. However we have a group of individuals known as agricultural extension workers who're akin to teachers for farmers. They assist in navigating the challenges brought about by changing weather patterns by sharing techniques such as diversifying crop choices and implementing water conservation methods.

While these heroes strive to make a difference they face obstacles such as constraints and regulatory restrictions that hinder farmers from adopting farming approaches. Nonetheless there is an opportunity for effort, among leaders, farmers and the wider community to lend support. By investing in educating farmers about farming methods we can effectively adapt to climate changes. Ensure an ample food supply for all. It's time for us to collaborate and prioritize the wellbeing of our farmers and the security of our food resources!

References

Adams, R. M., Hurd, B. H., Lenhart, S., & Leary, N. (1998). Effects of global climate change on agriculture: an interpretative review. Climate Research, 11(1), 19-30.

Alene, A. D., Zeller, M., & Pender, J. (2011). The impact of agricultural extension on adoption and use of new technologies: Evidence from the introduction of orange-fleshed sweet potatoes in Uganda. International Journal of Agricultural Sustainability, 9(1), 66-77.

Altalb, A. A. T., Filipek, T., & Skowron, P. (2015). The role of agricultural extension in the transfer and adoption of agricultural technologies. Asian Journal of Agriculture and Food Sciences, 3(5).

Anonymous. (2007), International Crops Research Institute for the Semi-Arid Tropics (ICRISAT). 'ICT-Enabled Knowledge Sharing in Support of Extension: Addressing the Agrarian Challenges of the Developing World Threatened by Climate Change, with a Case Study from India.' Open Access, 2007, oar.icrisat.org/2604/1/ICT-enabled_knowledge.pdf.

Barah, B. C., Ghimire, R. P., Subedi, D., Sharma, K. R., & Thapa, A. (2016). Understanding and improving local rice varieties by the means of rice knowledge systems in Nepal. Renewable Agriculture and Food Systems, 31(5), 477-490.

Best-Waldhober, M. D., Daamen, D., Ramirez, A., Faaij, A., Hendriks, C., & Visser, E. D. (2012). Informed public opinion in the Netherlands: Evaluation of CO_2 capture and storage technologies in comparison with other CO2 mitigation options. International Journal of Greenhouse Gas Control, 10, 169-180.

Boko, M., Niang, I., Nyong, A., Vogel, C., Githeko, A. & Tarhule, A. (2007). Africa. In M. L. Parry, O. F. Canziani, J. P. Palutikof, P. J. van der Linden, & C. E. Hanson (Eds.), Climate Change 2007: Impacts, Adaptation and Vulnerability. Contribution of Working Group II to the Fourth Assessment Report of the Intergovernmental Panel on Climate Change (pp. 433-467).

Braun, A. R., Jiggins, J., Röling, N., & Van den Berg, H. (2006). A global survey and review of farmer field school experiences. Rome: FAO.

Challinor, A. J., Watson, J., Lobell, D. B., Howden, S. M., Smith, D. R., & Chhetri, N. (2014). A meta-analysis of crop yield under climate change and adaptation. Nature Climate Change, 4(4), 287-291.

Dorward, P., Clarkson, G. and Stern, R. (2015). Participatory Integrated Climate Services for Agriculture (PICSA): field manual. A step-by-step guide to using PICSA with farmers. Walker Institute, University of Reading, pp64. ISBN 9780704915633.

Dwarakinath, R. (2006), Changing Tasks of Extension Education in Indian Agriculture. In A.W. Vanden Ban, and R. K. Samanta, (Eds.), Changing Roles of Agricultural Extension in Asian Nations. pp. 56-79. Delhi: B.R. Publishing, India.

Evans, E, B., David P Rowell, D, P, and Semazzi, H, M. (2020), The Future-Climate, Current-Policy Framework: Towards an Approach Linking Climate Science To Sector Policy Development. Barbara E Evans et al 2020 Environ. Res. Lett. 15. pp. 01-11.

Food and Agriculture Organization of the United Nations (FAO), (2018). The 10 elements of agroecology: Guiding the transition to sustainable food and agricultural systems.

Ghasemi, M., Piri, J., & Esmaeili, A. (2017). An overview on the irrigation systems and water management in the world. Advances in Plants & Agriculture Research, 7(1), 257-263.

Godfray, H. C. J., Beddington, J. R., Crute, I. R., Haddad, L., Lawrence, D., Muir, J. F., & Toulmin, C. (2010). Food security: The challenge of feeding 9 billion people. Science, 327(5967), 812-818.

Graham Clarkson , Peter Dorward , Sam Poskitt , Roger D. Stern , Dominic Nyirongo , Katiuscia Fara , John Mwangi Gathenya, Caroline G. Staub , Adrian Trotman, Gloriose Nsengiyumva , Francis Torgbor, Diana Giraldo. (2022), Stimulating small-scale farmer innovation and adaptation with Participatory Integrated Climate Services for Agriculture (PICSA): Lessons from successful implementation in Africa, Latin America, the Caribbean and South Asia. Elseiver: Climate Change. pp. 1-11.

Hansen, J. W., Mason, S. J., Sun, L., & Tall, A. (2011). Review of seasonal climate forecasting for agriculture in sub-Saharan Africa. Experimental Agriculture, 47(2), 205-240.

Jose, S. (2009). Agroforestry for ecosystem services and environmental benefits: An overview. Agroforestry Systems, 76(1), 1-10.

Kristjanson, P., Reid, R. S., Dickson, N., Clark, W. C., Romney, D., Puskur, R., & MacMillan, S. (2009). Linking international agricultural research knowledge with action for sustainable development. Proceedings of the National Academy of Sciences, 106(13), 5047-5052.

Lal, R. (2010). Beyond Copenhagen: mitigating climate change and achieving food security through soil carbon sequestration. Food Security, 2(2), 169-177.

Lal, R. (2015). Restoring soil quality to mitigate soil degradation. Sustainability, 7(5), 5875-5895.

Lin, B. B. (2011). Resilience in agriculture through crop diversification: Adaptive management for environmental change. BioScience, 61(3), 183-193.

Lobell, D. B., & Gourdji, S. M. (2012). The influence of climate change on global crop productivity. Plant Physiology, 160(4), 1686-1697.

Lobell, D. B., Bala, G., & Duffy, P. B. (2013). Biogeophysical impacts of land use on climate: Model simulations and observations. Journal of Climate, 26(19), 5422-5434.

Lobell, D. B., Schlenker, W., & Costa-Roberts, J. (2011). Climate trends and global crop production since 1980. Science, 333(6042), 616-620.

Meinzen-Dick, R., & Pradhan, R. (2002). Impacts of Agricultural Extension on Crop Production. Food Policy, 27(3), 265-277.

Minasny, B., Malone, B. P., McBratney, A. B., Angers, D. A., Arrouays, D., Chambers, A. & Hartemink, A. E. (2017). Soil carbon 4 per mille. Geoderma, 292, 59-86.

Msuya, C.P., Annor-Frempong, F.K., Magheni, M.N., Agunga, R., Igodan, C., Ladele, A., Huhela, K., Tselaesele, N.M., Msatilomo, H., Chomo, C. & Zwane, E., (2017). The role of agricultural extension in Africa's development: The importance of extension workers and the need for change. Int. J. Agric. Ext., 5(1):51-58.

Muzari, W., Mutsvangwa, E. P., Thierfelder, C., Ficarelli, P. P., Rusinamhodzi, L., Ngwira, A., & Shapiro, B. (2017). Smartphone-based digital tools for agricultural extension in Zimbabwe. International Journal of Education and Development using Information and Communication Technology, 13(1), 4.

Nair, P. K. R., Nair, V. D., & Kumar, B. M. (2009). Agroforestry as a strategy for carbon sequestration. Journal of Plant Nutrition and Soil Science, 172(1), 10-23.

Ogle, S. M., Swan, A., Paustian, K., & Ellert, B. H. (2012). No-till management impacts on CH4 and N2O fluxes from agricultural soils in western Canada. Soil and Tillage Research, 121, 87-94.

Oseni, M. O., & Hinson, R. (2014). Web-based extension services and agricultural productivity in Sub-Saharan Africa. Information Development, 30(2), 115-128.

Osgood, D. E., Ben-Horin, J., Dalton, M., & Kachergis, E. (2018). Climate-informed agricultural insurance for cereal farmers in Senegal. Climate Risk Management, 19, 61-69.

Pannell, D. J., Marshall, G. R., Barr, N., Curtis, A., Vanclay, F., & Wilkinson, R. (2006). Understanding and promoting adoption of conservation practices by rural landholders. Australian Journal of Experimental Agriculture, 46(11), 1407-1424.

Pretty, J., & Bharucha, Z. P. (2014). Sustainable intensification in agricultural systems. Annals of Botany, 114(8), 1571-1596.

Reicosky, D. C., & Kemper, W. D. (2015). Long-term no-till in the northern Great Plains. Journal of Soil and Water Conservation, 70(6), 125A-130A.

Rejesus, R. M., Palis, F. G., & Rodriguez, D. (2011). Crop diversification and risk management in the face of climate change: Evidence from the Philippines. Journal of Agricultural and Applied Economics, 43(3), 343-361.

Rezaei, M., Shahbazi, H., Hassani, A., Zavadskas, E. K., & Tamošaitienė, J. (2015). A multi-objective programming for environmentally sustainable agricultural production: A case study on paddy fields in Amol County, Iran. Journal of Environmental Management, 157, 157-166.

Smith, P., Martino, D., Cai, Z., Gwary, D., Janzen, H., Kumar, P. & Pan, G. (2008). Greenhouse gas mitigation in agriculture. Philosophical Transactions of the Royal Society B: Biological Sciences, 363(1492), 789-813.

Steward, D. R., Bruggeman, A., Prasad, P. V., & Staggenborg, S. A. (2013). Climate change and drought: A risk assessment of crop-yield impacts. Climate Research, 57(3), 221-227.

Sulaiman, V, R and Hall, A. (2006), Extension Policy Analysis in Asian Nations. In Vanden Ban, A. W. and Samanta, R. K. (eds) Changing Roles of Agricultural Extension in Asian Nations. B. R. Publishing Corporation, India.

Thornton, P. K., Boone, R. B., & Galvin, K. A. (2005). Climate change and the pastoralist agenda: Impacts, opportunities, and challenges. Policy Matters, 14, 49-60.

Van den Ban, A. W., & Hawkins, H. S. (1996). Agricultural extension. Blackwell Science. - Douthwaite, B., & Maatman, A. (2016). A framework for mapping innovation by researching the context in which it occurs: A participatory research approach. Knowledge Management for Development Journal, 11(2), 50-65.

Vermeulen, S. J., & Campbell, B. M. (2010). Climate change and food systems. Annual Review of Environment and Resources, 35, 225-226.

Zhang, Y., Yan, L., & Xie, Y. (2011). Potential assessment on the renewable energy development and greenhouse gas emission reduction in China. Renewable Energy, 36(7), 1863-1870.

Zwaan, B. V., Dalla Longa, F., Boer, H. D., Johnson, F., Johnson, O., Klaveren, M. V., Wanjiru, H. (2018). An Elicitation of Public Acceptance of Renewable Energy in Kenya. Nairobi, Kenya: Deliverable 2.5 of the TRANSrisk Project (Horizon 2020).